Transactions on Intelligent Welding Manufacturing

Editors-in-Chief

Shanben Chen
Shanghai Jiao Tong University
PRC

Yuming Zhang
University of Kentucky
USA

Zhili Feng
Oak Ridge National Laboratory
USA

Honorary Editors

G. Cook, USA
K. L. Moore, USA
Ji-Luan Pan, PRC

S. A. David, USA
S. J. Na, KOR
Lin Wu, PRC

Y. Hirata, JAP
J. Norrish, AUS

T. Lienert, USA
T. J. Tarn, USA

Guest Editors

H. P. Chen, USA
J. C. Feng, PRC
H. J. Li, AUS

X. Q. Chen, NZL
D. Hong, USA
W. Zhou, SGP

D. Du, PRC
X. D. Jiao, PRC

D. Fan, PRC
I. Lopez-Juarez, MEX

Regional Editors

Asia: L. X. Zhang, PRC
America: Y. K. Liu, USA

Australia: Z. X. Pan, AUS
Europe: S. Konovalov, RUS

Associate Editors

Q. X. Cao, PRC
B. H. Chang, PRC
J. Chen, USA
H. B. Chen, PRC
S. J. Chen, PRC
X. Z. Chen, PRC
A.-K. Christiansson, SWE
Z. G. Li, PRC
X. M. Hua, PRC

Y. Huang, USA
S. Konovalov, RUS
W. H. Li, PRC
X. R. Li, USA
Y. K. Liu, USA
L. M. Liu, PRC
H. Lu, PRC
Z. Luo, PRC
G. H. Ma, PRC

Pedro Neto, PRT
G. Panoutsos, UK
Z. X. Pan, AUS
X. D. Peng, NL
Y. Shi, PRC
J. Wu, USA
J. X. Xue, PRC
L. J. Yang, PRC
M. Wang, PRC

S. Wang, PRC
X. W. Wang, PRC
Z. Z. Wang, PRC
G. J. Zhang, PRC
H. Zhang, B, PRC
H. Zhang, N, PRC
L. X. Zhang, PRC
W. J. Zhang, USA

Academic Assistant Editors

J. Cao, PRC
B. Chen, PRC
Y. Luo, PRC
N. Lv, PRC
F. Li, PRC

S. B. Lin, PRC
Y. Shao, USA
Y. Tao, PRC
J. J. Wang, PRC
H. Y. Wang, PRC

S. L. Wang, PRC
J. Xiao, PRC
J. J. Xu, PRC
Y. L. Xu, PRC
C. Yu, PRC

H. W. Yu, PRC
K. Zhang, PRC
W. Z. Zhang, PRC
Z. F. Zhang, PRC

Editorial Staff

Executive Editor (Manuscript and Publication):
Responsible Editors (Academic and Technical):

Dr. Yan Zhang, PRC
Dr. Na Lv, PRC
Dr. Jing Wu, USA

More information about this series at http://www.springer.com/series/15698

Shanben Chen · Yuming Zhang · Zhili Feng
Editors

Transactions on Intelligent Welding Manufacturing

Volume II No. 3 2018

Editors
Shanben Chen
Shanghai Jiao Tong University
Shanghai, China

Zhili Feng
Oak Ridge National Laboratory
Oak Ridge, TN, USA

Yuming Zhang
Department of Electrical
and Computer Engineering
University of Kentucky
Lexington, KY, USA

ISSN 2520-8519   ISSN 2520-8527   (electronic)
Transactions on Intelligent Welding Manufacturing
ISBN 978-981-13-7420-3   ISBN 978-981-13-7418-0   (eBook)
https://doi.org/10.1007/978-981-13-7418-0

Library of Congress Control Number: 2019936002

© Springer Nature Singapore Pte Ltd. 2020
This work is subject to copyright. All rights are reserved by the Publisher, whether the whole or part of the material is concerned, specifically the rights of translation, reprinting, reuse of illustrations, recitation, broadcasting, reproduction on microfilms or in any other physical way, and transmission or information storage and retrieval, electronic adaptation, computer software, or by similar or dissimilar methodology now known or hereafter developed.
The use of general descriptive names, registered names, trademarks, service marks, etc. in this publication does not imply, even in the absence of a specific statement, that such names are exempt from the relevant protective laws and regulations and therefore free for general use.
The publisher, the authors and the editors are safe to assume that the advice and information in this book are believed to be true and accurate at the date of publication. Neither the publisher nor the authors or the editors give a warranty, expressed or implied, with respect to the material contained herein or for any errors or omissions that may have been made. The publisher remains neutral with regard to jurisdictional claims in published maps and institutional affiliations.

This Springer imprint is published by the registered company Springer Nature Singapore Pte Ltd.
The registered company address is: 152 Beach Road, #21-01/04 Gateway East, Singapore 189721, Singapore

Editorial

This issue of the Transactions on Intelligent Welding Manufacturing (TIWM) is a collection of high-quality papers selected from "2018 International Conference on Robotic Welding, Intelligence and Automation (RWIA'2018)," December 7–10, 2018, Guangzhou, China. They include two Feature Articles, eight full Research Papers, and one Short Papers and Technical Notes contributing to intelligent welding manufacturing through understanding, sensing, and control of welding manufacturing processes.

The first Feature Article entitled "Modeling and Optimization of Adjustment of Human Welder on Weld Pool Dynamics for Intelligent Robot Welding" is contributed by a joint research team from Lanzhou University of Technology and University of Kentucky. An improved machine–human cooperative control system was developed in this article to obtain sufficient data pairs for modeling welder's adjustment on weld pool dynamics by the data-driven approach. The effect of welding torch orientation on weld pool dynamics was numerically studied for understanding the mechanism of human welder's adjustment and providing the key data to optimize the control system. A gray multiple linear regression model (GMLRM) was employed to analyze the contribution and interactive compensation of each adjusted parameter on the weld widths as the welding current randomly changes in a given range. A nonlinear adaptive kernel radial basis function neural network (AK-RBFNN) was also proposed to improve the model accuracy.

The second Feature Article entitled "Progress and Trend in Intelligent Sensing and Control of Weld Pool in Arc Welding Process" is contributed by researchers from Lanzhou University of Technology. In this article, the current progress in sensing of arc welding pool is detailed and challenges in measuring the weld pool are analyzed. The key factor that hinders the development of intelligent robotic welding is identified, and approaches that realize intelligent welding are also discussed. Lastly, the trend of intelligent welding manufacturing is predicted.

The first Research Paper "Spectral Analysis of the Plasma Emission During Laser Welding of Galvanized Steel with Fiber Laser" is contributed by researchers from Harbin Institute of Technology. Spectral information during the laser welding of galvanized steel was obtained. The plasma spectra were analyzed under different laser powers and sheet gaps. Signal filter and statistical method were used to process the obtained spectral information. It was found that the plasma temperature increases with the laser power, and plasma temperature and spectral intensity demonstrate a minimum with the sheet gap. Statistical process control method was used to analyze the relationship between the welding quality and the spectral information. It was found that the welding defects, in laser welding of galvanized steel, are detectible from the spectra.

The second Research Paper "A Method for Detecting Central Coordinates of Girth Welds Based on Inverse Compositional AAM in Tube-Tube Sheet Welding" studies detecting central coordinates of girth welds based on inverse compositional AAM by using a triaxial flaw detection device for the tubular heat exchanger. It is contributed by researchers from Shanghai Jiao Tong University and involves the design of software, calibration algorithm, center detection algorithm, etc. The accuracy of the developed algorithm was experimentally verified.

Research Paper "Spectral Signal Analysis Using VMD in Pulsed GTAW Process of 5A06 Al Alloy" is also contributed by researchers from Shanghai Jiao Tong University. It proposes an automatic discriminant criterion based on correlation coefficient to eliminate redundant wavelength signals in spectral domain (200–1100 nm), resulting in only a few spectral lines for further subsequent processing. To overcome the limit of EMD, the variational mode decomposition (VMD) algorithm is used to decompose the spectral signal into the determined number of intrinsic signals with fewer modal aliasing in the time domain.

Following Research Paper "Nonlinear Identification of Weld Penetration Control System in Pulsed Gas Metal Arc Welding" is a contribution from researchers at Tianjin University. The paper established a single-input, single-output (SISO) weld penetration control system in pulsed gas metal arc welding (GMAW-P). According to the nonlinear relationship between base current and arc voltage difference, a Hammerstein model with residual, composed of the nonlinear static model and linear dynamic model, was proposed to describe the nonlinear control system.

Research Paper "Effect of Transverse Ultrasonic Vibration on MIG Welded Joint Microstructure and Microhardness of Galvanized Steel Sheet" is contributed by joint researchers from School of Mechanical Engineering and School of Environment and Chemical Engineering, Nanchang University. A comparison test of conventional MIG welding and ultrasonic-MIG hybrid welding was carried out in this paper. The effects of transverse ultrasonic vibration on weld formation, weld microhardness, and weld microstructures during ultrasonic-MIG hybrid welding of 1-mm-thick galvanized steel sheet were discussed.

Following Research is entitled "Effect of Scanning Mode on Microstructure and Physical Property of Copper Joint Fabricated by Electron Beam Welding" contributed by authors from Nanchang Hangkong University. Copper T2 thin sheet with a thickness of 2 mm was vacuum electron beam welded. Scanning electron microscopy, optical microscopy, microhardness tester, and tensile testing machine were used to facilitate the investigation. The formation of the weld surface by different scanning methods was also studied.

Editorial

The seventh Research entitled "Research on Virtual Reality Monitoring Technology of Tele-operation Welding Robot" is contributed by researchers from Beijing Institute of Petrochemical Technology. In this paper, various aspects of the tele-operation of welding robot were discussed, such as robot construction, 3D model building and application in Unity, network communication between the virtual robot and the physical robot, and tele-operation welding test trial.

The last Research Paper is "The Influence of Powder Layers Intervention on the Microstructure and Property in Brazing Joints of Titanium/Steel," by researchers from Lanzhou University of Technology. In this paper, comparison tests, where different types of metal powder layers were used in the brazing joints of titanium/steel, were carried out. Mechanical properties and microstructures of the joints were also analyzed against processing parameters and different powders.

The Short Paper "Process Research on Diode Laser-TIG Hybrid Overlaying Welding Process" is a contribution from researchers at Lanzhou University of Technology. A laser-TIG overlaying welding method is proposed to improve forming quality and reduce equipment cost. An experiment system was set up to study the influence of welding parameters on forming characteristics.

I am confident that the above papers in this TIWM issue significantly contribute to the frontier of intelligent welding manufacturing, as well as the topics related to the conference RWIA'2018.

Yuming Zhang, Ph.D.
TIWM Editor-in-Chief
yuming.zhang@uky.edu

James R. Boyd
Professor of Electrical Engineering, University of Kentucky
Fellow, American Welding Society (AWS)
Fellow, American Society of Mechanical Engineers (ASME)
Fellow, Society of Manufacturing Engineers (SME)

Contents

Feature Articles

Modeling and Optimization of Adjustment of Human Welder on Weld Pool Dynamics for Intelligent Robot Welding 3
Gang Zhang, Yukang Liu, Yu Shi, Ding Fan and Yuming Zhang

Progress and Trend in Intelligent Sensing and Control of Weld Pool in Arc Welding Process 27
Ding Fan, Gang Zhang, Yu Shi and Ming Zhu

Research Papers

Spectral Analysis of the Plasma Emission During Laser Welding of Galvanized Steel with Fiber Laser 47
Bo Chen, Zhiwei Chen, Han Cheng, Caiwang Tan and Jicai Feng

A Method for Detecting Central Coordinates of Girth Welds Based on Inverse Compositional AAM in Tube-Tube Sheet Welding 65
Yu Ge, Yanling Xu, Huanwei Yu, Chao Chen and Shanben Chen

Spectral Signal Analysis Using VMD in Pulsed GTAW Process of 5A06 Al Alloy 83
Haiping Chen, Gang Li, Na Lv and Shanben Chen

Nonlinear Identification of Weld Penetration Control System in Pulsed Gas Metal Arc Welding 95
Wandong Wang, Zhijiang Wang, Shengsun Hu, Yue Cao and Shuangyang Zou

Effect of Transverse Ultrasonic Vibration on MIG Welded Joint Microstructure and Microhardness of Galvanized Steel Sheet 109
Guohong Ma, Xiaokang Yu, Jian Li and Yinshui He

Effect of Scanning Mode on Microstructure and Physical Property of Copper Joint Fabricated by Electron Beam Welding............. 119
Ziyang Zhang, Shanlin Wang, Jijun Xin, Yuhua Chen and Yongde Huang

Research on Virtual Reality Monitoring Technology of Tele-operation Welding Robot...................................... 133
Canfeng Zhou, Long Wang, Yu Luo, Hui Gao, Juan Li and Guoxue Gao

The Influence of Powder Layers Intervention on the Microstructure and Property in Brazing Joints of Titanium/Steel 147
Pengxian Zhang, Yibin Pang and Shilong Li

Short Papers and Technical Notes

Process Research on Diode Laser-TIG Hybrid Overlaying Welding Process.. 161
Ming Zhu, Buyun Yan, Xubin Li, Yu Shi and Ding Fan

Information for Authors...................................... 169

Author Index... 171

Feature Articles

Modeling and Optimization of Adjustment of Human Welder on Weld Pool Dynamics for Intelligent Robot Welding

Gang Zhang, Yukang Liu, Yu Shi, Ding Fan and Yuming Zhang

Abstract An improved machine–human cooperative control system was developed to obtain sufficient data pairs for modeling welder's adjustment on weld pool dynamics by data-driven approaches. Spectral analysis shows that weld widths are apparently changed due to low-frequency variations of the welder's hand movement, which can be filtered by a low-pass filter to remove the high-frequency components. The effect of welding torch orientation on weld pool dynamics was numerically studied to understand the mechanism of human welder's adjustment and to provide the useful data for control system optimization. A gay multiple linear regression model (GMLRM) was employed to analyze the contribution and interactive compensation of each adjusted parameter on the weld widths as the welding current randomly changes in a given range. A nonlinear adaptive kernel radial basis function neural network (AK-RBFNN) was also proposed to improve the model accuracy. Results indicate that the redundant, coupled, and integrated hand adjustments are adopted to maintain the desired weld pool status, and the human welder's adjustment reflect nonlinear, complex characteristics. Results also show that the proposed AK-RBFNN model can appraise the weld widths with a good accuracy.

Keywords GTAW · Weld pool dynamics · Numerical simulation · Multiple linear regression

---

G. Zhang (✉) · Y. Shi · D. Fan
State Key Laboratory of Advanced Processing and Recycling Non-Ferrous Metals, Lanzhou University of Technology, Lanzhou 730050, China
e-mail: zhanggang@lut.cn

G. Zhang · Y. Liu · Y. Zhang
Electrical and Computer Engineering Department, Institute of Sustainable Manufacturing, University of Kentucky, Lexington, KY 40506, USA

© Springer Nature Singapore Pte Ltd. 2019
S. Chen et al. (eds.), *Transactions on Intelligent Welding Manufacturing*, Transactions on Intelligent Welding Manufacturing, https://doi.org/10.1007/978-981-13-7418-0_1

1 Introduction

Automation and robotic welding technology have been widely applied in automobile manufacture to increase the production and reduce the labor cost. In high-quality and precision-required nuclear, pressure vessels manufacture, etc., however, manual operations are still employed to guarantee the weld bead quality due to the complex and varying conditions. During manual gas tungsten arc welding (GTAW) process, skilled welders can estimate the welding results, mainly the weld joint penetration and weld defects, through their observations on the weld pool, and they can also adjust the welding parameters including welding current, arc length, torch speed, and orientation intelligently to control the welding process to obtain the desired weld bead. However, the bodily limitations such as fatigue, stress, and changeable emotion might degrade the welder's capabilities in daily operations, cause long-time health risk, and inversely affect the weld quality. Hence, modeling and transmitting their experience and skills to the automated welding machine can significantly overcome the human welder's physical shortages and improve the welding precision.

Extensive studies have been performed on learning human welder operations using various methods. Byrd et al. used a virtual reality (VR) simulator to assess novice welders for ensuring high-quality welding, but this method could not distinguish between experienced and trained novice welders [1]. Hashimoto measured the skilled welder's arm movement using six bends and electromyography sensors and trained the new welder's operation [2]. Seto et al. [3] studied the decision-making process of the expert welders in aluminum alloy welding using analytic hierarchy process approach. Asai et al. [4] developed a visualized and digitized system to display the welder's operation and comparatively analyze his/her welding skills. Hashimoto also applied four cameras to capture the 3D positions of torch electrode tip and then detailed the welder's skill levels and responses on the varying weld pool surface, identifying the difference between skilled and unskilled welders [5]. These studies indicate that certain human welder operations have been recorded and learned. However, the human welder's adjustment on the varying weld pool has not been parameterized and modeled.

With the limit of the author's searching capability, limited investigations of modeling the welder's responses on the weld pool are recently reported. Zhang et al. firstly proposed an innovative vision-based sensing method to successfully measure 3D weld pool surface by its width, length, and convexity in GTAW process [6] and then linearly modeled the dynamic behavior of the human welder adjusting the weld pool status to achieve a desired weld penetration [7, 8]. Comparing with the real industry and actual stochastic characteristics of the welder, it is found that linear model may not be sufficient to describe intrinsic nonlinear and fuzzy inference features of the welder. Hence, Liu et al. constructed a neuro-fuzzy-based human intelligence model and implemented it as an intelligent controller in automated GTAW process to obtain a consistent weld penetration [9–11]. An adaptive neuro-fuzzy inference system (ANFIS) was further applied to model the welding current/speed adjusted by the welder to the characteristic parameters of weld pool with a higher accuracy [12, 13].

The previous established models only reflect the effect of welding speed/current adjustment on the weld pool status and partly learn the human welder's adjustment behavior. They cannot clearly explain the essential reasons why the human welder can avoid and control weld defects occurred in welding process by adjusting the torch orientations or arc length or welding speed or all of them. The compensated mechanism of all the adjusted parameters on the stable welding process remains unclear. This paper developed an improved machine–human cooperative control system to conduct teleoperation experiments and obtain the data pairs. Based on the analysis of characteristics of human welder's adjustment, the thermal-physical welding process guiding to the human welder's adjustments was numerically studied for optimizing the control system and adjusted algorithm of welding torch orientation. Furthermore, a gay multiple linear regression (GMLR) and an adaptive kernel RBF neural network (AK-RBFNN) system was proposed to model the correlation between the weld widths (front-side and back-side weld width) and the human welder's adjustments, respectively. The interactive compensation of all parameters was analyzed. The AK-RBFNN model was demonstrated by experiments.

2 Materials and Experimental Procedures

The machine–human cooperative teleoperated control system is briefly described, which is shown in Fig. 1 together with the experimental setup developed by Liu and Zhang [14].

This system includes two workstations: welding station and virtual station. In the virtual station shown in Fig. 1c, a human welder can observe the mock-up where the workpiece geometry is displayed and moves the virtual welding torch accordingly as if he/she actually operates it in front of the workpiece. The human welder adjustment is accurately measured by a leap motion sensor, and the obtained 3D coordinates of the virtual welding torch tip are sent to the PC. The welding station includes an industrial welding robot, eye view camera, and a compact 3D weld pool surface sensing system designed by Zhang et al. [15]. The robot utilized in this study is Universal Robot UR-5 with 6 DOF, which is connected with the PC via Ethernet using TCP/IP protocol and socket programming. The welding torch is equipped to the robot arm and follows the arm movement of the human welder using the commands received from the PC. Figure 1a illustrates a detailed view of the 3D weld pool sensing system. Camera 2 captures the weld pool image and sends it back to the PC. A virtual signal (arrow with direction and amplitude) is added to the image for the welder to view (a sample image is shown in an upper right). A real-time 3D weld pool surface constructed system was also developed to provide institutive information of the weld pool for the human welder to make a smart decision [15]. A typical 3D weld pool surface image is shown in a lower right of Fig. 1a. Figure 1d depicts a detailed view of the human hand movement tracking system. A projector is used to project the virtual image (weld

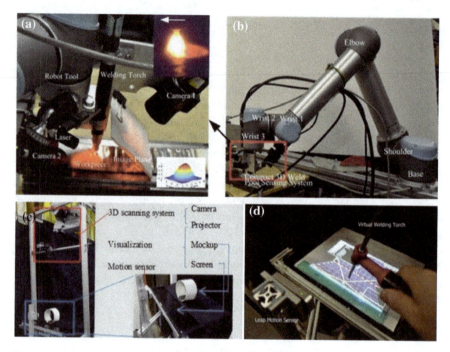

Fig. 1 Human–machine cooperative teleoperated virtual welding system: **a** real welding station; **b** detailed view of the compact sensing system; **c** virtual welding system; **d** view of human hand movement tracking system

Table 1 Chemical compositions in wt% of 304 stainless steel

Element	C	Mn	P	S	Si	Cr	Ni	Fe
Value	≤0.08	≤2.00	≤0.045	≤0.03	≤1.00	18.0–20.0	8.0–11.0	Balance

pool image captured by camera 2) for the welder to see. The direction and amplitude of the arrow are computed by the arm movement controller. The human welder then moves the virtual welding torch accordingly. Leap motion sensor accurately captured the human welder's hand movement, which can track fingers or similar items to a special precision of 0.01 mm, which is detailed in Liu et al. [16].

The direct current electrode negative GTAW was applied to conduct experiments on the 304 stainless steel pipe, and the chemical compositions of the pipe are listed in Table 1. The outer diameter and wall thickness of the pipe are 113.5 and 3 mm, respectively.

During the welding process, the stainless steel pipe kept stationary, and the welding torch moved along the circumference of the pipe. Because the human welder hand movement includes the both deterministic and stochastic characteristics, to analyze the effect of this action on the weld pool status and inversely study the human welder response on the varying weld pool, eleven experiments were conducted where the

Table 2 Experimental parameters

Welding parameters			
Current (A)	Welding speed (mm/s)	Arc length (mm)	Argon flow rate (L/min)
40–48	0.6–1.6	3–7	12
Sensing parameters			
Project angle (°)	Laser to weld pool distance (mm)	Imaging plane to weld pool distance (mm)	
31.5	25	101	
Camera parameters			
Shutter speed (ms)	Frame rate (fps)	Camera to imaging plane distance (mm)	
2	30	57.8	

human welder moved the virtual welding torch along the mock-up, and his/her hand motion coordinates recorded by leap sensor were sent to the robot for following and completing the welding task. In these experiments, welding current is randomly changed from 40 to 48 A resulting in a fluctuated weld pool. Other experimental parameters are shown in Table 2. The sampling frequency in this study is 3 Hz due to the relatively slow torch adjustment of the welder.

3 Results and Discussion

3.1 Experiment Data and Spectral Analysis

To investigate the human welder responses on the welding process, four parameters containing welding speed S, arc length l, torch rotation angle along the welding direction R_x, and perpendicular to the welding direction R_y are proposed to characterize the human welder's hand adjustment, and four weld pool characteristic parameters including front-side and back-side width, length and surface convexity are used to represent the welding process, which are considered as the major source a human welder perceives to complete the welding experiments. Particularly, weld penetration reflected by the back-side width closely correlates with the change of the pool surface convexity, which has been demonstrated by Zhang et al. [17]. Eleven experiments performed with Table 2 parameters are plotted in Fig. 2.

Figure 2 shows that the front-side width fluctuates with the change of the arc length, orientation R_x and R_y as well as welding speed; the tendency of the pool surface convexity is roughly opposite to the welding current and arc length, which also correlates to the torch orientation R_x and R_y. The pool length is positive to the rotation R_y and the welding speed. When the welding current increases suddenly, the heat input to the weld pool from arc increases accordingly, and then, an enlarged weld pool is observed by the welder, and he/she tries to reasonably add the torch

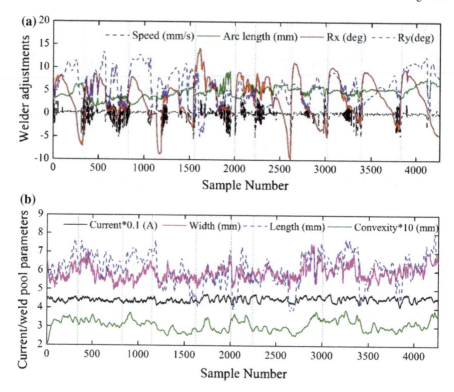

Fig. 2 Data pairs measured in eleven experiments: **a** welder hand adjustments; **b** weld pool characteristic parameters

moving speed to reduce the heat flux for maintaining the thermal equilibrium of the weld pool. In the case of novice human welder, a large welding speed may be adopted for lack of welding experience and fuzzy estimation of the welding process, such that a narrow width and shallow weld penetration would be produced, even resulting in some weld defects (i.e., undercutting and humping). Skilled human welder, to maintain the consistent and smooth weld width and weld joint penetration, tries to decrease the arc length to increase the arc drag force and adjust the rotation angle R_x and R_y as the welding speed increases. These dynamic adjustments of human welder include the nonlinear and fuzzy features because of the welder's intrinsic fuzzy inference behavior. In the next two sections, linear and nonlinear models are established to elaborate this complex behavior.

To thoroughly analyze the nonlinear characteristics of the human welder's movements and accurately model them with the weld pool status, the histograms of the adjusted parameters and pool parameters in eleven experiments are plotted in Fig. 3, respectively.

Figure 3 shows that 80% of the welding speed and the distance of the torch electrode tip to the pipe surface range from −0.5 to 1.6 mm/s and 3 to 6 mm,

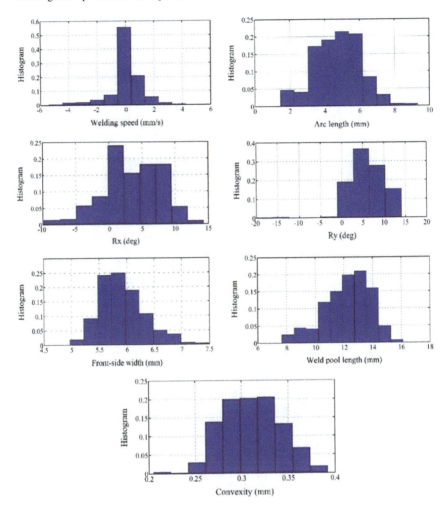

Fig. 3 Histograms of the welder adjustments and weld pool parameters

respectively; the rotation angle R_x ranges from $-5°$ to $10°$, and the R_y covers the range $[0°, 14°]$. There is 90% of the front-side width, length, and surface convexity of the weld pool varying from 5.3 to 6.5 mm, 10–14.5 mm, and 0.27–0.35 mm, respectively. Some high frequency or large variations occur in the human welder's adjustments (i.e., sample number 600–800 and 1800–1900 in the welding speed; sample number 1550–1700 and 2100–2400 in the R_x). However, a well-proportioned front-side width is obtained, and a relative consistent weld joint penetration may be produced based upon the histogram distribution of the weld pool convexity where the maximum alteration achieves to 0.08 mm. The reasons are inferred that, based on the past learned experience, the human welder can primarily observe the current state of the weld pool and accurately estimate its trend that would be changed next time, and

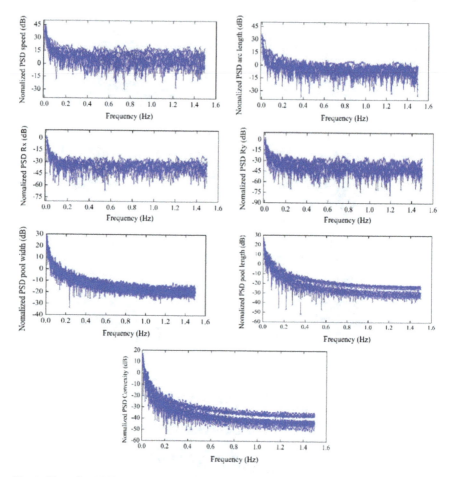

Fig. 4 Normalized PSDs for human welder adjustments and weld pool parameters

then, he/she can adjust other parameters (i.e., l, R_x or R_y) to compensate the large variations of the pre-altered parameter (i.e., welding current, welding speed). This compensation can thus decrease the large variations in the weld pool and produce a relative smooth weld bead.

Spectral analysis-based approach is employed to study the effect of high-frequency adjustments on the stable weld pool, and power spectral density (PSD) of the welding speed, arc length, R_x and R_y are computed and plotted in Fig. 4, respectively.

Figure 4 shows that the maximum power of the signals including the welding speed, arc length, R_x and R_y is generated at 0.78, 0.39, 0.52, and 1.2 Hz, respectively. The frequency distribution of each of the parameters includes high frequencies (above 0.75 Hz). The signal corresponded to the high frequency may be generated by the surrounding random noises or the improper hand execution of the human

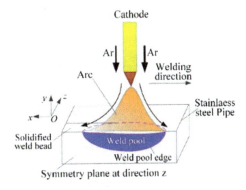

Fig. 5 Schematic diagram of the 3D weld pool

welder. Generally, these high frequencies included in the signal are considered as a disturbance and are not needed to be followed by the automatic robot. In addition, comparative analysis of the measured weld pool parameters to the hand adjustment parameters shown in Fig. 3 indicates that the high-frequency variations in the welder adjustments weakly affect the front-side weld width and pool surface convexity. That is to say, the welding process has no active feedback to the high-frequency adjustments. The welding process can thus be considered as a low-pass filter. Hence, the high frequency or large variations are not needed to be significantly concerned when the human welder responses on the varying weld pool are modeled.

3.2 Numerical Modeling and Analysis

Analysis of human welder's adjustment mentioned in Sect. 3.1 shows that the weld pool system is an inertia system which has a time delay for the change of human welder. Thereby, precisely obtaining the delay time is critical for accurately control of weld pool and optimizing the control system. To this end, a simple mathematical model of the 3D weld pool with taking the traveling speed into account was established based on the reasonable assumptions proposed by Du et al. [18] and Wang et al. [19]. Because all the experiments in this study are completed using a low welding current (40–48 A), and studies indicate that the weld pool surface deformation under low current (<100 A) little affects the fluid flow, temperature distribution, and geometry of the weld pool, which is verified by Rokhlin and Guu [20]. Hence, the small deformation of the weld pool surface is neglected, and the schematic diagram of the weld pool is shown in Fig. 5.

(1) Conservation equations:

Mass continuum equation

$$\frac{\partial \rho}{\partial t} + \nabla \cdot (\rho \boldsymbol{v}) = 0 \tag{1}$$

Momentum conservation equation:

$$\frac{\partial (\rho \boldsymbol{v})}{\partial t} + \nabla \cdot (\rho \boldsymbol{v}\boldsymbol{v}) = -\nabla \cdot P + \nabla \cdot \tau + j \times B + \rho g + S_u \tag{2}$$

Energy conservation equation:

$$\frac{\partial (\rho c_p T)}{\partial t} + \nabla \cdot (\rho c_p T \boldsymbol{v}) = \nabla \cdot (k \nabla T) + S \tag{3}$$

In the above equations, t is simulated time, ρ is density, v is velocity vector, T is temperature, c_p is specific heat and k is thermal conductivity, P is pressure, and τ is viscous shear tensor, and is given by

$$\begin{cases} \tau_{ij} = \mu \left(2\frac{\partial v_i}{\partial x_i} - \frac{2}{3} \nabla \cdot \boldsymbol{v} \right), & i = j \\ \tau_{ij} = \mu \left(\frac{\partial v_i}{\partial x_j} + \frac{\partial v_j}{\partial x_i} \right), & i \neq j \end{cases} \tag{4}$$

where v_i and x_i are the components of the velocity and position in either the $i = x$, y, or z direction, respectively, and μ is dynamic viscosity coefficient.

An enthalpy-porosity technique proposed by Voller and Prakash [21] is adopted to describe the melting process of the 304 stainless steel pipe. S_u represents a momentum source to characterize the flow in the mushy zone, which can be written as:

$$S_u = A v = -C \frac{(1 - f_1)^2}{f_1^3 + B} v \tag{5}$$

where C is a relatively large constant (1.6×10^3), B is a relatively small constant (0.001), f_1 is the liquid fraction and addressed with the same method used by Wang et al. [19].

As a source term of the energy equation, latent heat melting solid material is needed to be considered to describe the melting process. The source term can be expressed as following.

$$S = S_h = \frac{\partial (\rho f_1 L)}{\partial t} + \nabla \cdot (\rho \boldsymbol{v} f_1 L) \tag{6}$$

where L is the latent heat of fusion for the anode material, and f_1 is liquid fraction.

(2) Boundary conditions.

The net heat transfer input at the top weld pool surface can be calculated using the below equation. In Eq. (7), q_{arc} is the arc energy. q_{rad}, q_{conv}, and q_{evap} are the radiation, convection, and evaporation part, respectively. All of them can be computed by the equations described by Goldak [22].

$$q = q_{arc} - q_{conv} - q_{rad} - q_{evap} \quad (7)$$

At the symmetric and the other-side surfaces, the heat flux density can be expressed by following equation, respectively.

$$\frac{\partial T}{\partial Z} = 0 \quad (8)$$

$$q = -q_{conv} - q_{rad} - q_{evap} \quad (9)$$

For the heat source, a Gaussian thermal flux distribution was used in the case of torch orientation in accordance with the normal orientation of the workpiece. In the case that the torch orientation is not perpendicular to the normal direction of the workpiece, an elliptical heat source was used to calculate the heat input. Both of them can be written as:

$$\begin{cases} q_{Gaussian} = \frac{\eta U I}{2\pi r_H^2} \exp\left(\frac{-r^2}{2r_H^2}\right) \\ r^2 = (x - vt)^2 + y^2 \end{cases} \quad (10)$$

$$q_{elliptical} = \frac{\eta U I}{2\pi ab} \exp\left(\frac{-(x-vt)^2}{m^2} - \frac{y^2}{n^2}\right) \quad (11)$$

where η is arc power coefficient; r_H^2 is the radius of arc heat and is set as 1.75 mm for 4 mm arc length and as 1.42 mm for 2 mm arc length according to the results obtained by Tsai and Eagar [23]; m, n represents semi-major axial length and the semi-minor axial length of the ellipse, respectively; in this study, we set $m = 1.98$ mm; $n = 1.54$ mm according to melting geometry of the weld pool surface; V and t present welding speed and welding time, respectively; U and I are arc voltage and welding current.

The temperature-dependent coefficient of the surface tension can be calculated using following equations listed by Wu [24].

$$\gamma = \gamma^0 - A_c(T - T_0) - R_g T \Gamma_g \ln\left[1 + k_1 a_g \exp(-\Delta H^0 / R_g T)\right] \quad (12)$$

$$\frac{\partial \gamma}{\partial T} = -A_c - R_g \Gamma_g \ln(1 + K_{ag} a_g) - \frac{K_{ag} a_g \Delta H^0 \Gamma_g}{T(1 + K_{ag} a_g)} \quad (13)$$

Fig. 6 Schematic diagram of the computation domain

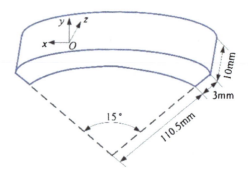

where A_c is a constant; γ^0 is the surface tension of the pure metal at temperature T^0; R_g, $\Gamma_g a_g$, ΔH^0, and k_1 are gas constant, supersaturated parameter, activity coefficient of element S in material, standard absorption heat, and separation enthalpy, respectively; K_{ag} represents the constant coefficient of separation. They can be found in Ref. [24].

The electromagnetic force in the weld pool can be calculated by Eq. (14). Here, H represents the wall thickness of the pipe. σ_j is the current flux radius and equals the r_H.

$$\begin{cases} F_x = -\frac{\mu_m I^2}{4\pi^2 \sigma_j^2 r} \exp\left(\frac{r^2}{2\sigma_j^2}\right)\left[1 - \exp\left(-\frac{r^2}{2\sigma_j^2}\right)\right]\left(1 - \frac{z}{H}\right)^2 \frac{x}{r} \\ F_y = -\frac{\mu_m I^2}{4\pi^2 \sigma_j^2 r} \exp\left(\frac{r^2}{2\sigma_j^2}\right)\left[1 - \exp\left(-\frac{r^2}{2\sigma_j^2}\right)\right]\left(1 - \frac{z}{H}\right)^2 \frac{y}{r} \\ F_y = -\frac{\mu_m I^2}{4\pi^2 H r^2}\left[1 - \exp\left(-\frac{r^2}{2\sigma_j^2}\right)\right]\left(1 - \frac{z}{H}\right) \\ r^2 = (x - vt)^2 + y^2 \end{cases} \quad (14)$$

(3) Numerical considerations.

The thermo-physical properties parameters of the argon employed by Boulos et al. [25] were used in this study. A given computation domain shown in Fig. 6 was selected, and the corresponding temperature, fluid flow, and weld pool geometry were calculated, respectively. For the workpiece, a non-uniform grid point system was utilized with finer grid sizes near the weld pool area to improve the computation accuracy. The weight of element S in base material was set as 0.022% (220 ppm).

The conservation equations were resolved, discretized, and solved iteratively by the segregated solver of the CFD commercial FLUENT code, enhanced with dedicated UDF to handle the source terms and extra scalar equations needed for the electromagnetic variables. The standard SIMPLE algorithm was used for the force–velocity coupling, and a second-order interpolation scheme was employed in all the equations. The corresponding parameters used in the model are listed in Table 3. In addition, the convection heat transfer coefficient h_c, thermal conductivity k, and viscosity μ were processed using the same approach proposed by Wu [24].

Table 3 Physical parameters used in the numerical model calculation

Parameters	Value
Radiative emissivity ε_r (K^{-1})	0.4
Stefan–Boltzmann constant σ (W m^{-2} K^{-4})	5.367×10^{-8}
Latent heat L (J kg^{-1})	2.47×10^{-5}
Ambient temperature T_0 (K)	300
Permeability μ_m (H m^{-1})	$4\pi \times 10^{-7}$
Solidus temperature T_s (K)	1673
Liquid temperature T_l (K)	1727
Density ρ (kg m^{-3})	7200
Specific heat c_p (J kg^{-1} K^{-1})	753
Arc heat coefficient η	0.65

At instant $t = 0$, $T = T0$, and $v = 0$. In simulation, the welding parameters change independently, and the step time of them is set at the 5 s after starting welding.

(4) Computational results.

Based on the established model, the simulated results are given in Fig. 7, which illustrate the temperature and velocity field in the weld pool before and after the step time.

Figure 7a shows the computational results at welding current 45 A, arc length 4 mm, welding speed 1 mm/s, torch tilt angel 0° (torch orientation is agreed to the normal direction of the workpiece all the time). It is clear that there exist inward flows caused by the positive surface tension temperature gradient ($\partial \gamma / \partial T > 0$) which mainly depends on the content of active elements in base material and significantly changes the weld joint penetration.

Figure 7b depicts the temperature and velocity field in the weld pool only when the arc length changes from 2 to 4 mm. Comparisons of Fig. 7a and b show that the weld penetration, top width and length of weld pool was increased. The peak temperature of the weld pool also increases from 1895.44 to 1905.42 K, and this results from more concentrated arc power due to the increase in the arc voltage. On the other hand, analysis of the Eq. (10) clearly shows that the heat input to the weld pool increases as the arc length becomes longer. Hence, this makes sense that the weld joint penetration deepens, and the top width and length of the weld pool changes wider and longer due to the expansion of the anode heating area covered by the arc column.

Comparative analysis of Fig. 7a, c implies that the pipe penetration clearly increases when the welding speed descends from 1 to 0.5 mm/s; the top width of the weld pool increases 1 mm (5 mm under 1 mm/s, 6 mm under 0.5 mm/s), and the weld pool length decreases 0.5 mm (5.5 mm at 1 mm/s, 5 mm/s at 0.5 mm/s); the maximum inward flow rate of liquid metal reduces 0.01 m/s, and the maximum temperature of the weld pool changes from 1905.42 to 1912.84 K. These variations can be contributed to the increase in the heat input per unit length for the pipe

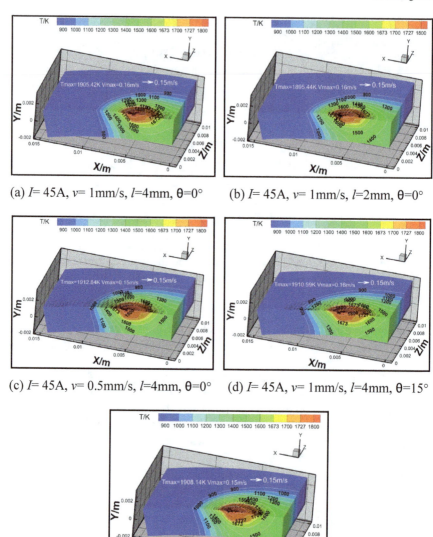

Fig. 7 Temperature and velocity fields under different welding parameters

through analyzing the heat source Eq. (10). The inward flow sufficiently leads the high-temperature liquid metal to the bottom of the weld pool at lower welding speed and melts more solid metal to increase the weld joint penetration. The decrease in the welding speed further prevents the backflow of the liquid metal from the head of weld pool to its tail, which in turn reduces the weld pool length. However, it is clear that the variation in the welding speed changes the weld joint penetration more apparently than the weld pool width and length.

Figures 7a, d illustrate that when the torch angle changes from 90° to 75° (their included angle 15°), the weld pool geometry alters less obviously, and the maximum temperature has a small change about 5 K. Only the weld pool becomes spindly, which reflects that the heat distribution on the weld pool surface changes; the heat input to the weld pool maintains the same across the welding direction. When the torch angle achieves to 75°, the arc center will move to the ahead or tail of the weld pool, resulting in a lower temperature gradient which, in turn, weakens the strength of the Marangoni convection on the pool surface. Analysis of Eq. (14) points out that the torch angle variation further decreases the electromagnetic force due to the increase in the anode heating radius and then weakens the strength of inward convection induced by the electromagnetic force. Besides, the buoyancy force on the head or tail of the weld pool increases because of the movement of the arc center, which accelerates the outward Marangoni convection. However, investigations show that the electromagnetic and buoyancy forces change the weld pool shape smaller than the surface tension on the weld pool surface [26]. It is easily understood that a very small change occurs in the weld shape, especially, weld width, and depth, because of the small variation on the heat distribution under torch angle 75°. The similar conclusions are obtained in stationary GTAW by Parvez et al. [27]. However, experiments show that the varying torch orientation can improve the weld bead appearance and avoid some weld defects [28].

Figure 7e illustrates the temperature and velocity field in the weld pool with larger welding current 48 A compared with Fig. 7a. It is clear that the weld width becomes little narrow and the pipe penetration increases, and an inward flow in the weld pool is formed. With the increase in the welding current, the heat input to the weld pool also increases, and then, more heat input is transferred to the bottom of the weld pool by the inward flow pattern. Such that, more solid metal is melted, and a deeper weld bead is achieved.

When the welding torch orientation changes from 90° to 75°, the width of weld pool becomes narrower, and the length of weld pool is much longer. After 0.4 s, the unstable weld pool system achieves a new stable status. However, the weld penetration is slightly increased, after 0.6 s, the changes of depth remain a constant which is shown in Fig. 8. Thus, the response time of control system of weld pool can be predicted via above simulated results. When the performed time interval of controller can be set 0.6 s for satisfying the delay time, so that, the period of seeking an appropriate control time via vast repeated experiments is immensely shortened.

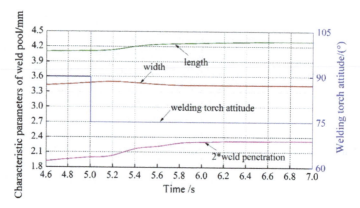

Fig. 8 The variation of weld pool parameters with change of welding torch orientation

3.3 Data-Driven Modeling

A general model structure correlating human welder hand adjustments and weld widths is written as follows:

$$\begin{cases} \widehat{W}_t(k) = f_1\big(S(k-1), l(k-1), R_x(k-1), R_y(k-1)\big) \\ \widehat{W}_b(k) = f_2\big(S(k-1), l(k-1), R_x(k-1), R_y(k-1)\big) \end{cases} \quad (15)$$

where $\widehat{W}_t(k)$ and $\widehat{W}_b(k)$ represent the front-side and back-side weld width estimated from the model f_1 and f_2 at instant k, respectively; $S(k-1)$, $l(k-1)$, $R_x(k-1)$, and $R_y(k-1)$ are the measured welding speed, arc length, rotation angle along and perpendicular to the welding direction at instant $k-1$, respectively.

Two criteria including average error and root mean square error are proposed to appraise the performance of the identified models. The model average error is defined by Eq. (16).

$$E_{\text{ave}} = \frac{1}{n} \sum_{k=1}^{n} |\widehat{W}_{t,b}(k) - W_{t,b}(k)|, \quad (k = 1, \ldots, n) \quad (16)$$

The root mean square error (RMSE) is calculated by the following equation.

$$\text{RMSE} = \sqrt{\sum_{k=1}^{n} \big(\widehat{W}_{t,b}(k) - W_{t,b}(k)\big)^2 / n} \quad (17)$$

where n is the number of data pairs; $W_{t,b}(k)$ represents the measured front-side and back-side weld width at instant k, respectively.

3.3.1 Gray Multiple Linear Regression Model (GMLRM)

In GMLRM, a new time series values are first obtained using the conventional GM (1, 1) model detailed in Ref. [29], and then, the correlation of the independent variables and dependent variables is modeled by multiple linear regression method. Such that the trend of the dependent variables can be predicted with a good accuracy. GMLRM can also overcome the multi-collinearity shortage of the normal MLRM mode which is detailed in Ref. [30].

The following model is established and identified using standard least squares algorithm:

$$\begin{cases} \widehat{W}_t(k) = \eta_s S(k-1) + \eta_l l(k-1) + \eta_x R_x(k-1) + \eta_y R_y(k-1) + C_t \\ \widehat{W}_b(k) = \alpha_s S(k-1) + \alpha_l l(k-1) + \alpha_x R_x(k-1) + \alpha_y R_y(k-1) + C_b \end{cases} \quad (18)$$

where η, α, and C are the model parameters.

The identified linear model is expressed by following equation.

$$\begin{cases} \widehat{W}_t(k) = 0.0086 S(k-1) + 0.0471 l(k-1) - 0.0216 R_x(k-1) \\ \qquad + 0.0053 R_y(k-1) + 5.741 \\ \widehat{W}_b(k) = -0.005 S(k-1) + 0.0571 l(k-1) + 0.008 R_x(k-1) \\ \qquad + 0.008 R_y(k-1) + 2.771 \end{cases} \quad (19)$$

Equation (19) linearly illustrates the correlation between the welder's adjustments and the front-side and back-side weld widths. The static gains for the welding speed, arc length, R_x and R_y are (0.0086, −0.005), (0.047, 0.057), (−0.0216, 0.008), and (0.0053, 0.008), respectively. They reveal that the front-side and back-side weld widths tend to increase as the value of the parameters weighted the positive coefficient increases, and to decrease with the reduction in the parameters weighted the negative coefficient. This makes sense because the increases in the arc length and rotation angle R_y indicate that the arc radius covering the weld pool surface increases, and more metal is melted such that the front-side weld width is enlarged. However, the heat input to the weld pool reduces because the enlarged arc radius makes more heat loss into the atmosphere, such that, the weld joint penetration becomes shallow. Fortunately, the increase in the welding speed can compensate this change of the weld penetration and the front-side weld width, which can be reflected by the static gains (−0.005 and 0.0086), respectively. Besides, the long arc length easily makes the arc drifted to the pool surface and produces an uneven weld bead shape, and some probable weld defects. Hence, the skilled welder takes a smart action that the torch orientations are adjusted in accordance with the varying arc length for maintaining a consistent weld bead width and weld penetration. The compensation of the torch orientation adjustments can be reflected by the weighting coefficients of the R_x and R_y. The sign of the coefficients is determined by its contribution on the whole change of the weld pool. Comparative analysis further implies that

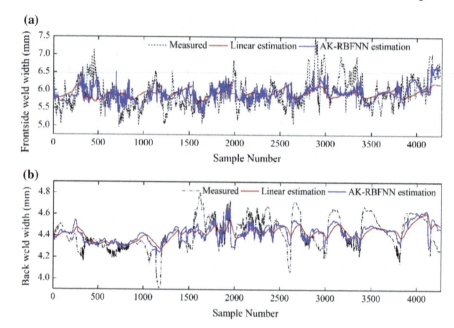

Fig. 9 Linear and AK-RBFNN modeling results: **a** front-side width and **b** back-side width

when one of the input parameter is given, the other three inputted parameters can be changed randomly to fit the desired top weld width and weld joint penetration. When the two inputted parameters are ensured simultaneously, only two numerical solutions can fit the Eq. (19). It is well known that the range of all the parameters adjusted by a skilled welder is generally confirmed rather than randomly changed. It is impossible that each of the inputted parameter is precisely controlled by the welder in the welding process due to the combined influence of the welder physical shortages with the varying welding conditions. Hence, the welder's hand adjustment presents a redundant characteristic, and holistic and comprehensive result of controlling all the parameters. These actions are not a simple linear correlation between the inputted parameters but a complex and fuzzy decision process. The linear results are plotted in Fig. 9.

Clearly, the trend of the weld widths can be estimated with an acceptable accuracy as the inputted parameters change. The average model error and RMSE are listed in Table 4. Figure 9 depicts that some details in the front-side and back-side weld width cannot be estimated/tracked well (i.e., sample number from 400 to 500 in Fig. 9a and sample number from 2100 to 2500 in Fig. 9b). Substantial static fitting errors are frequently plotted.

Table 4 Model error comparisons

Output	Model type	RMSE (mm)	Average error (mm)
W_t	GMLRM model	0.3929	0.3096
	AK-RBFNN model	0.3406	0.2581
W_b	GMLRM model	0.1271	0.1014
	AK-RBFNN model	0.1067	0.0828

3.3.2 AK-RBFNN Model

A normal structure of RBFNN generally includes three layers: an input layer, a nonlinear hidden layer, and a linear output layer. The mapping of the input and the output is given as:

$$y = \sum_{i=1}^{m_1} w_i \emptyset_i(\|x - x_i\|) + b \quad (20)$$

where m_1 represents the amount of the neurons in the hidden layer; x_i is the centers of the RBF network; w_i represents the weight coefficient connecting the hidden layer to output neuron; b is the error term of the output neuron, and $\emptyset_i$ is the basis function of the ith hidden neuron. A single output neuron is considered in this work for modeling the front-side and back-side width of weld pool, respectively. The Gaussian kernel is employed due to its versatility, which can be written as following [31]:

$$\emptyset_i(\|x - x_i\|) = \exp\left(\frac{\|-x - x_i^2\|}{\sigma^2}\right) \quad (21)$$

where σ is the spread of the Gaussian kernel. Recently, the cosine metric is also used to measure the distance due to its complimentary properties, which is shown in Eq. (22).

$$\emptyset_{i1}(x.x_i) = \frac{x.x_i}{\|x\|\|x_i\| + \gamma} \quad (22)$$

where the term γ is a very small positive constant. Hence, the basis function can be expressed by fusing the cosine $\emptyset_{i1}(x.x_i)$ and Euclidean $\emptyset_i(\|x - x_i\|)$ distances [32].

$$\emptyset_i(x, x_i) = \alpha_1 \emptyset_{i1}(x.x_i) + \alpha_2 \emptyset_i(\|x - x_i\|) \quad (23)$$

where α_1 and α_2 are the fusion weights. Consequently, the overall mapping, at the nth learning interaction linked to a specific time, can be written as:

$$y(n) = \sum_{i=1}^{m_1} w_i(n) \emptyset_i(\boldsymbol{x}, \boldsymbol{x}_i) + b(n) \tag{24}$$

where the synaptic weights $w_i(n)$ and bias $b(n)$ are adapted at each iteration.
In order to calculate the adapted rule, a cost function $\varepsilon(n)$ is defined as:

$$\varepsilon(n) = 0.5(d(n) - y(n))^2 \tag{25}$$

where $d(n)$ is the desired output at the nth iteration, and $y(n)$ is the actual output of neuron. The update rule for the kernel's weight is given by:

$$\Delta \alpha(n) = -\eta \frac{\partial \varepsilon(n)}{\partial \alpha(n)} \tag{26}$$

The update equations of the weight and bias, based on the chain rule of differentiation for the cost function combing Eq. (26), are given as:

$$w_i(n+1) = w_i(n) + \eta \varepsilon(n) \emptyset_i(\mathbf{x}, \mathbf{x}_i) \tag{27}$$

$$b_i(n+1) = b_i(n) + \eta \varepsilon(n) \tag{28}$$

Hence, according Eqs. (27) and (28), the weights for the participating kernels are dynamically updated without prior assignment.

To further improve the model accuracy, nonlinear AK-RBFNN method is employed to estimate the front-side width and the weld penetration based on the inputted parameters denoted as the input of the neurons. The predicted results are shown in Fig. 9, and the errors are listed in Table 4. It is apparently observed that the model errors are largely improved by incorporating the nonlinear correlation between the inputs and the outputs. However, the physical mechanism that the system inputs adjust its outputs is not clear instead of becoming fuzzier than the linear model. Hence, based on the thermal-physical theory in the welding process, the welder's adjustment behavior is clearly elaborated by numerical analysis.

3.4 AK-RBFNN-Model Verification and Analysis

Despite the consistency of the human welder's performance, data-driven model should have the similar performance to estimate the current weld pool status in other experiments. To verify the accuracy of the AK-RBFNN model, verification experiments were conducted and one of them was selected to carefully analyze. The corresponding characteristic parameters were captured and shown in Fig. 10a–d, respectively.

Modeling and Optimization of Adjustment of Human Welder ...

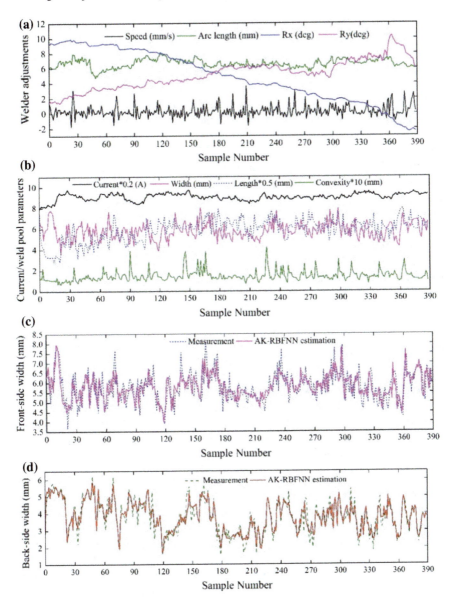

Fig. 10 Verification experiment results

As Fig. 10 shows, the AK-RBFNN model can estimate the front-side and back-side width of the weld pool with acceptable accuracy. It is noticed, however, that certain weld pool widths are not well predicted by this model. (Sample number from 60 to 70, 160 to 170 and 290 to 300 in front-side width; Sample number from 30 to 40, 260 to 275 and 310 to 320 in back-side width), yet this model estimated front-side and back-side width of the weld pool is maintained at 5.5 and 3.5 mm, respectively. Careful analysis of sample number 150–170 indicates that when the welding current decreases from 48 to 45 A, the weld penetration first decreases from 5 mm to 4 mm, and the top weld width becomes narrower (6–5.2 mm). However, the welder observes the smaller back-side width of the weld pool, inferring that the heat input to the weld pool decreases and then tries to reduce the welding speed from 1.3 to 0.2 mm/s to increase the heat flux on the unit volume. Meanwhile, he/she raises the arc length to produce more arc heat for increasing heat input. Although these adjustments of welder increase the heat input of weld pool and penetrate the workpiece, the relatively large adjustment of welding parameters fluctuates the weld pool, and a fine tuning, to ensure the smooth and consistent weld width and weld penetration, is completed for the welding speed, arc length, and torch orientations. The whole process of welder adjustments represents the redundant characteristics.

4 Conclusion

In this paper, the human welder's hand adjustments in accordance with the varying weld pool are recorded by an innovative machine–human cooperative teleoperated welding system and are analyzed using statistical method. The correlation between the welder's adjustment parameters and the weld widths are modeled and verified by experimental data, and further elaborated in detail by numerical modeling. The following conclusions can be drawn.

1. An improved machine–human cooperative control system is employed to obtain sufficient data pairs (welder's hand adjustment and weld pool parameters) for modeling the correlation of them.
2. Spectral analysis of the data pairs indicates that high-frequency variations in the welder's adjustments weakly affects the front-side weld width and back-side width. The low-frequency components, on the contrary, have a large effect on the variation of the weld pool status. The weld pool system can thus be considered as a low-pass filter, and the high-frequency components can be negligible.
3. Linear GMLRM and nonlinear AK-RBFNN models correlating the welder's hand adjustment to the weld widths are identified by the data pairs, and careful analysis indicates that the human welder's hand movement includes a redundant characteristic and thus an interactive compensated mechanism can be utilized to illustrate that the consistent and symmetrical weld bead appearance is obtained in all the varying inputted parameters.

4. The numerical results imply that the rich responses/skills of the welder are more reflected in the control of the various forces acting on the weld pool and the energy distribution inputted from the arc.
5. Verification experiments demonstrate the accuracy of the established AK-RBFNN model with lower average error and RMSE.

Acknowledgements This work is funded by the Scientific research project of university of GanSu Province (2018A-018) and Hongliu Outstanding Young Talents Support Project of Lanzhou University of Technology. The authors would like to thank the assistance from Xinxin WANG and Lei XIAO on the numerical model establishment.

References

1. Byrd AP, Stone RT, Anderson et al (2015) The use of virtual welding simulators to evaluate experienced welders. Weld J 12:389s–395s
2. Hashimoto N (2015) Measurement of welder's movement for welding skill analysis. Bull Hiroshiima Inst Tech Res 49:83–87
3. Seto N, Mori K, Hirose S (2012) Extracting of the skill from expert welders for aluminium alloy and investigation of their viewpoints using analytic hierarchy process. Light Metal Weld 8:14–22
4. Asai S, Ogawa T, Takebayashi H (2012) Visualization and digitization of welder skill for education and training. Weld World 56:26–34
5. Hashimoto N (2015) Difference of improving welder's skill through training progression. Bull Hiroshima Inst Tech Res 49:75–81
6. Zhang WJ, Liu YK, Zhang YM (2012) Characterization of three-dimensional weld pool surface in gas tungsten arc welding. Weld J 91:195s–203s
7. Zhang WJ, Zhang YM (2012) Modeling of human welder response to 3D weld pool surface: Part1-Principles. Weld J 91:310s–318s
8. Zhang WJ, Zhang YM (2012) Modeling of human welder response to 3D weld pool surface: part 2-results and analysis. Weld J 91:329s–337s
9. Liu YK, Zhang WJ, Zhang YM (2015) Dynamic neuro-fuzzy based human intelligence modeling and control in GTAW. IEEE T Autom Sci Eng 12:324–335
10. Liu YK, Zhang WJ, Zhang YM (2013) Control of human arm movement in machine-human cooperative welding process. Control Eng Pract 21:1469–1480
11. Liu YK, Zhang YM (2014) Model-based predictive control of weld penetration in gas tungsten arc welding. IEEE T Control Syst T 22:955–966
12. Liu YK, Zhang WJ, Zhang YM (2013) Adaptive neuro-fuzzy inference system (ANFIS) modeling of human welder's responses to 3D weld pool surface in GTAW. J Manuf Sci Eng T ASME 135:0210101–02101011
13. Liu YK, Zhang YM (2015) Iterative local ANFIS-based human welder intelligence modeling and control in pipe GTAW process: a data-driven approach. IEEE T Mech 20:1079–1088
14. Liu YK, Zhang YM (2015) Controlling 3D weld pool surface by adjusting welding speed. Weld J 94:125s–134s
15. Zhang WJ, Zhang YM (2013) Analytical real-time measurement of three-dimensional specular weld pool surface. Meas Sci Technol 24:115011–115029
16. Liu YK, Shao Z, Zhang YM (2014) Learning human welder movement in pipe GTAW: a virtualized welding approach. Weld J 93:388s–398s
17. Zhang G, Shi Y, Li CK et al (2014) Research on the correlation between the status of three-dimensional weld pool surface and weld penetration in TIG welding. Acta Metall Sin 8:995–1002

18. Du HY, Wei YH, Wang WX et al (2009) Numerical simulation of temperature and fluid in GTAW-arc under changing process conditions. J Mater Process Tech 209:3725–3765
19. Wang XX, Fan D, Huang JK et al (2014) A unified model of coupled arc plasma and weld pool for double electrodes TIG welding. J Phys D Appl Phys 47:275202–275211
20. Rokhlin SI, Guu AC (1993) A study of arc force, pool depression, and weld penetration during gas tungsten arc welding. Weld J 8:382s–390s
21. Voller VR, Prakash C (1987) A fixed grid numerical modeling methodology for convection diffusion mushy region phase-change problems. Int J Heat Mass Trans 30:1709–1718
22. Goldak J (1984) A new finite element model for welding heat sources. Metall Trans B 15:299–305
23. Tsai MS, Eagar TW (1985) Distribution of the heat and current fluxes in gas tungsten arcs. Metall Trans B 16:841–846
24. Wu CS (2008) Welding thermal process and molten pool dynamic. Mechanical Industry Publication, Beijing
25. Boulos IM, Fauchais P, Pfender E (1994) Thermal plasma-fundamentals and applications. Plenum 1:388
26. Mougenot J, Gonzalez JJ, Freton P et al (2013) Plasma–weld pool interaction in tungsten inert-gas configuration. J Phys D Appl Phys 46:135206–135220
27. Parvez S, Abid M, Nash DH et al (2013) Effect of torch angel on arc properties and weld pool shape in stationary GTAW. J Eng Mech 139:1268–1277
28. Ancona A, Lugarà PM, Ottonelli F et al (2004) A sensing torch for on-line monitoring of the gas tungsten arc welding process of steel pipes. Meas Sci Technol 15:2412–2422
29. Wu LF (2016) Using fractional GM (1,1) model to predict the life of complex equipment, Grey Syst. Theory Appl 6:32–40
30. Prieto M, Tanner P, Andrade C (2016) Multiple linear regression model for the assessment of bond strength in corroded and non-corroded steel bars in structural concrete. Mater Struct 49:4749–4763
31. Wettschereck D, Dietterich T (1992) Improving the performance of radial basis function networks by learning center locations. In: Advances in neural information processing systems vol 4, pp 1133–1140
32. Aftab W, Moinuddin M, Shaikh MS (2014) A novel kernel for RBF based neural networks. Abstr Appl Anal 2014:1–10

Dr. Gang Zhang was born in 1986 and works at Lanzhou University of Technology now as a lecturer, who is interested in the fields included monitoring and control of weld penetration in arc welding process, dissimilar metals joining with advanced welding methods, and wire additive arc manufacturing. He as a visiting scholar has studied in Welding and manufacturing center of University of Kentucky, USA in 2016 and has published 20 research papers which were accepted by Welding Journal, Journal of Materials Processing Technology, Robotics and Computer-Integrated Manufacturing and so on.

Progress and Trend in Intelligent Sensing and Control of Weld Pool in Arc Welding Process

Ding Fan, Gang Zhang, Yu Shi and Ming Zhu

Abstract Intelligent robotic welding is a trend in the development of welding manufacturing and has a great application prospective. Weld pool dynamics plays a vital role in assuring the production of high-quality welds. Precision measurement of the weld pool surface characteristics is a bottleneck for accurate control of weld penetration as well as for successful development of next-generation intelligent welding system. In this article, the current progress in arc welding pool sensing is detailed and challenges in weld pool measurement analyzed. The key factor that hinders the development of intelligent robotic welding systems is identified and the approaches that realize the intelligent welding are also discussed. Lastly, the trend of intelligent welding manufacturing is predicted.

Keywords Weld pool · Intelligent sensing and control · Human–machine cooperation · Deep learning

1 Introduction

Welding named industrial sewing, as a traditional materials processing technology, plays a vital role in modern manufacturing. It has been applied in many industrial fields, such as aerospace, shipbuilding, and ocean drilling. Due to the difference in individual human welder's experience and skills, weld qualities vary, affecting the weld production and manufacturing. As advanced manufacturing technologies develop, intelligent robotic welding becomes urgent. However, welding robots are currently still programmed. As such, welding parameters and welding path are pre-

---

D. Fan (✉) · Y. Shi · M. Zhu
School of Materials Science and Engineering, Lanzhou University of Technology, Lanzhou 730050, China
e-mail: fand@lut.cn

D. Fan · G. Zhang · Y. Shi
State Key Laboratory of Advanced Processing and Recycling Non-Ferrous Metals, Lanzhou University of Technology, Lanzhou 730050, China

© Springer Nature Singapore Pte Ltd. 2019
S. Chen et al. (eds.), *Transactions on Intelligent Welding Manufacturing*,
Transactions on Intelligent Welding Manufacturing,
https://doi.org/10.1007/978-981-13-7418-0_2

specified, and the weld quality must be ensured by strict control of welding conditions [1–3]. Although this may hinder the occurrence of some weld defects, other factors including thermal conduction, materials composition, and weld deformation may still be subject to fluctuations and can comprise the results produced by welding robots. On the other hand, the adaptive adjustment and control of weld pool dynamics in manual arc welding by highly skilled human welders may sometimes reduce some weld defects as response to the varied conditions. Unfortunately, current welding robots are not equipped with those adaptive abilities to adapt to the complex welding structures and conditions [4]. Thereby, it is very urgent to develop advanced intelligent robotic welding technologies and intelligent welding system which can also adapt the changes and fluctuations that may occur during welding to better assure weld qualities and help address of the issue of increasing shortage of skilled human welders.

Weld pool reflects effects of the welding operation imposed human welders or welding robots on welding process and weld quality and thus contains useful information possibly including process stability, weld defects, and weld bead properties. However, how to extract potential information from it to accurately predict the weld quality and control the weld pool dynamics is a challenge and it is a long hard task to raise the intelligent level of robotic welding to a level comparable with skilled human welders. This paper detailedly expounds the research progress in the measurement of weld pool dynamics using advanced sensing technics and intelligent control of weld pool behavior in arc welding process. Moreover, the trend of intelligent robotic welding is also discussed.

2 Sensing and Measurement of Weld Pool

Weld pool dynamics closely correlates to the occurrence of weld defects, thermophysical process, and weld bead shape. Welding process involves multidisciplinary knowledge including arc physics, heat diffusion, metallurgy, and dynamics; the final status of weld pool was determined by the weld metal fluid flow properties, heat and mass transfer. Due to the influence of strong arc light, it is uneasy to acquire a clear weld pool image to precisely control welding process and weld quality. Investigations on the sensing and measurement of weld pool in arc welding process indicate there are two aspects. One is active vision-based sensing approaches which utilized the arc light to provide the light source of camera and then imaged the weld pool. This method has some advantages such as simple device (one camera and digital image acquisition card) and operation. Another is passive vision-based method which used one additional optical source like flashing light to illuminate the weld pool or cover the strong arc light for sampling the weld pool image using a high-speed camera. Although this method added the extra device, complicating the experimental system and operation, a higher and clearer weld pool image was obtained in welding process which can benefit to identify the stability of weld pool and control of weld quality.

2.1 Passive Vision-Based Sensing of Weld Pool

Nowadays, passive vision-based approaches, such as machine vision sensing [5–7], weld pool oscillation [8, 9], infrared sensing [10], and optical sensing [11], have been developed to measure the weld pool. Moreover, machine vision sensing method is widely applied owing to its simple devices, direct and non-contacting advantages. Based on machine vision sensing method, two-dimensional or three-dimensional information of weld pool was acquired to identify its characteristics and predict its weld penetration. Shanghai Jiao Tong University Professor Chen group [12–20] first analyzed the visual sensing system from the point of the view of light intensity and established the relationship between the gray image, arc length, and welding current, and the weld pool image at base current period in pulsed tungsten inert arc welding was sampled and the 2D information of weld pool was obtained, and a neural network between 2D parameters of weld pool and welding current was established for control of the top weld bead formation [13, 21, 22].

Richardson et al. [23] first built the vision-based sensing system and acquired the top surface of weld pool when the strong arc light was covered by W electrode. Because the limited area covered by electrode, the acquired weld pool image cannot reveal the edge of weld pool. A special imaging processing algorithm that includes denoising, sharpening, and lightness enhanced was developed to extract its characteristics. Filtering strong arc light is urgent to acquire clearer weld pool image. Thereby, researchers have proposed several filter technologies, such as arc spectrum-based composite filter and narrowband composite filter, which were used to acquire the 2D information in aluminum pulsed TIG welding or TIG welding [24, 25]. Moreover, Luo et al. [26] projected a high-power shutter structured laser as an active light source to the weld pool surface for suppressing the strong arc light, and the vision sensing system as well as weld pool image is shown in Fig. 1, respectively.

Although the strong arc light was suppressed with coupled filter technology and the relative clear image of weld pool was acquired, some certain characteristics of

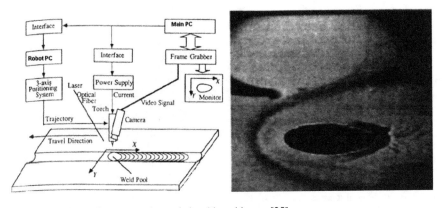

Fig. 1 Experimental system and sampled weld pool image [25]

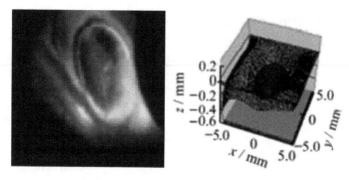

Fig. 2 3D weld pool reconstructed by SFS [27]

weld pool such as the width and length of weld pool are reflected. By observing the variation of weld pool, it is easily found that the weld pool includes 3D characteristics, especially, weld metal fluid flow closely associates with the weld formation and microstructures as well as mechanical properties.

Hence, it is very urgent that new approaches are developed to measure 3D characteristic of weld pool to control of weld pool dynamics and weld bead quality. There are three new methods to observe the weld pool, including shape from shielding, binocular stereo vision, and structured light vision sensing. Li et al. [27, 28] first analyzed the characteristics of aluminum GTAW process and then established the reflection map model of weld pool surface obtaining the height of weld pool surface. The reconstructed weld pool is shown in Fig. 2.

A large division between the reconstructed weld pool surface and the practical weld pool is produced because of the influences of mirror-like reflection of weld pool surface and the ideal assumptions of computational model included lambert surface and far away light sources. Mnich et al. [29] focused on the weld pool behavior of GMAW pipe welding and reconstructed the weld pool using binocular stereo vision approach, which is shown in Fig. 3.

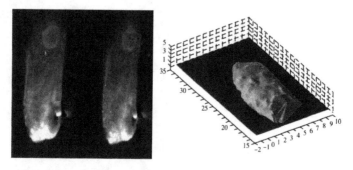

Fig. 3 3D weld pool reconstructed by binocular stereo vision method [29]

Progress and Trend in Intelligent Sensing and Control ...

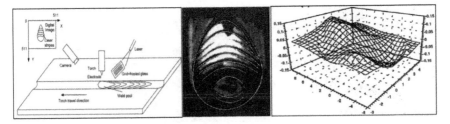

Fig. 4 3D weld pool reconstructed by structured laser light [30]

To adapt the rapid variation of weld pool, a higher synchronicity of the two cameras and high-quality image must be satisfied in this method. Moreover, a fast response of welding power needs to be provided for realizing the instant shutoff of welding arc.

2.2 Active Vision-Based Sensing of Weld Pool

University of Kentucky Professor Zhang et al. [30–32] first proposed a new approach that the structured illumination laser light was used to measure 3D weld pool surface. The measurement system was composed of high-power laser generator and high-speed camera sampling system. The acquired weld pool images are shown in Fig. 4.

To clearly identify 3D characteristics of weld pool surface, Go et al. [33] further improved the previous experiment system using the laser grating. The height of weld pool surface can be calculated by the deformation of laser grating. However, this method cannot accurately obtain the edge of weld pool and quantitatively characterize the weld pool. The laser grating-based experimental system and acquired weld pool image are shown in Fig. 5, respectively.

To accurately measure the 3D weld pool, Song and Zhang et al. [34, 35] projected a dot matrix structured laser light onto the whole weld pool surface, and high-resolution weld pool image was acquired by a high-speed camera, and then used the special imaging processing algorithm to extract the 3D characteristics of weld pool and

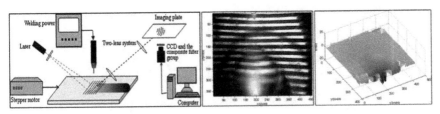

Fig. 5 Experimental system and 3D weld pool reconstructed by laser grating method [32]

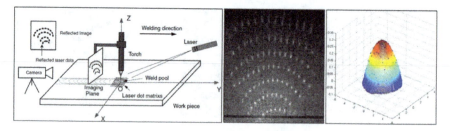

Fig. 6 Experimental system and 3D weld pool reconstructed by dot matrix laser light [34, 35]

reconstruct the 3D morphology. The experimental system acquired dot matrix laser image and reconstructed image is shown in Fig. 6, respectively.

The dot matrix-based measurement of weld pool proposed by Professor Zhang has been improved by Lanzhou University of Technology Zhang because of the limited application with low welding current and convex weld pool surface. Zhang [36, 37] improved the reconstructed algorithm and successfully rebuilt the 3D geometry of concave weld pool produced by high welding current based on the updated slope of the reflected laser dots and the mixed pattern between the projected laser dot and the reflected laser dot. Furthermore, the correlation of weld pool status and weld penetration was quantitatively established, which is shown in Fig. 7.

2.3 Fluid Flow Measurement of Weld Pool

Because of arc pressure, arc dynamical change, and strong arc light as well as the high-temperature radiation, it is very difficult to directly measure the weld metal flow pattern in real time. In recent years, particle tracer method and X-ray method are developed to investigate the dynamic behavior of weld metal flow in GTAW. Li et al. [38] demonstrated that the direction of Marangoni convection in TIG weld

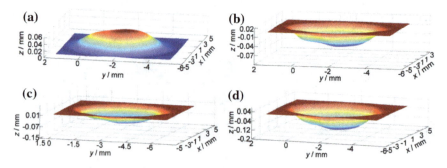

Fig. 7 Three-dimensional reconstructed shapes of weld pool surfaces [35]

pool was changed by oxygen content and the flow direction was measured by the pure silver. Oxide on the weld pool surface had undesirable impacts on the weld depth as the oxygen excessively accumulated in the weld pool. Naito et al. [39] used micro-focused high-speed X-ray transmission video system to visualize the liquid melt flow pattern in laser-arc welding. The specimens with W particles in the butt-joint interfaces were especially prepared for the observation of the liquid flow or convection in the molten pool. Experiments showed that the weld metal fluid flow presents an inward pattern. Unfortunately, only the partial information of weld metal flow is achieved by using above methods. Many physical quantities such as fluid flow speed, variable status, weld thermal cycles, and various solidification parameters cannot be directly measured in welding process due to the rapid solidification, severe changes of inward metal fluid. The investigations of weld metal flow behavior in GTAW are mainly centralized in the numerical simulation. [40–42] The weld pool surface deformation [43, 44], and weld pool temperature fields [45, 46], etc., have been simulated based on the established numerical mathematical model. In considering the effect of temperature and oxygen concentration on the surface tension, a 3D mathematical model was established to simulate the fluid flow patterns and weld penetration in activating flux tungsten inert gas welding (A-TIG) process. The results indicated that the change of the surface tension gradient in the molten pool was considered to be the principal mechanism for increasing penetration [47]. Yin et al. [48] simulated the weld metal flow in GTAW by establishing a coupled 3D numerical model between welding arc and weld pool with applied axial magnetic fields and obtained that the fluid of the center areas, in the circumferential direction, rotates in an opposite direction to that in the outer regions; in the axial direction, the fluid flows upwards at the center while downwards in the edge area of the weld pool. However, the simulation cannot precisely describe the variations of the weld pool and also cannot be verified by the experiments. Thereby, it is a very important and urgent task to develop a simple, rapid, and accurate method to measure the weld metal flow pattern or rate and characterize its dynamic behavior.

3 Weld Pool Based—Intelligent Control of Weld Quality

Weld pool dynamics is a very complex process which refers to metal melting, heat convection, solidification, and phase transformation. Furthermore, it is very difficult to establish a mathematical model between the weld pool behavior and the weld quality. Liu et al. [49] built a laser vision-based sensing system and extracted the image characteristics of weld pool in CO_2 welding process and then established neuro networks between the weld pool characteristic parameters and the welding current. This system realized the feedback control of weld pool by adjusting the welding current real time and improved the adaptive capacity of robot. Yue et al. [50] effectively fused the top width of weld pool with its temperature data based on multi-information fusion technology and established a BP neuron network to control the back width of weld bead in MIG process and obtain a continuous full

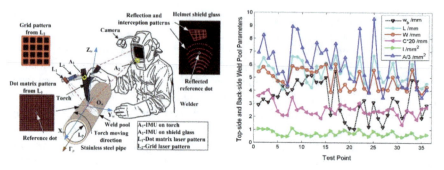

Fig. 8 3D characteristic parameters of weld pool [56]

penetration weld bead. Zhang et al. [51] first built the robust processing system for sensing the weld pool by means of machine-vision and then established the neuro networks model of the top and back parameters of weld pool, designing a differential controller assembled by self-learning neurons and proportion sum. At last, the back-side of weld pool was controlled by the developed system. Shi [52] also developed the flexible robot arc manufacturing system and the PID/weighted fuzzy controller to control of the aluminum pulsed MIG welding process. Experimental results demonstrated that the multi-parameters coupled MIG process can be controlled well by the developed control system. Du [53] adopted the nonlinear Hammerstein model to characterize the thermal welding process and developed a parameter preset feed-forward fuzzy-neuron networks controller. The welding current and welding speed was adjusted by combing the human welder experience and weld pool image for compensating the effect of welding assembly error, thermal deformation on the weld formation. Based on the multi-information fusion technology, LUT professor Fan et al. [54] established a fusion model of multi-characteristic signal including vision sensing information, welding current and voltage, arc acoustics to real time identify the status of weld pool and the stability of arc dynamics and metal transfer in aluminum alloy pulsed MIG welding process. Furthermore, a double-variable decoupling control system was proposed for GMAW-P of aluminum, and the experiments were performed, showing that good weld bead shape and stable welding process can be obtained through the double-variable decoupling control scheme without complex metal transfer control and considerable trial and error to identify suitable combinations of welding parameters in GMAW-P [55]. Current investigations have not fully considered the influence of human welder skills and responses on the weld quality.

University of Kentucky Zhang et al. [56] improved the previous laser vision-based sensing system and accurately obtained the characteristics of weld pool as well as described the human welder's responses and skills using certain defined physical parameters, which are shown in Fig. 8.

Based on this experimental system, the varying current experiments were conducted, and the human welder's response and the characteristics of weld pool were sampled, respectively. A linear data-driven model of welding current and

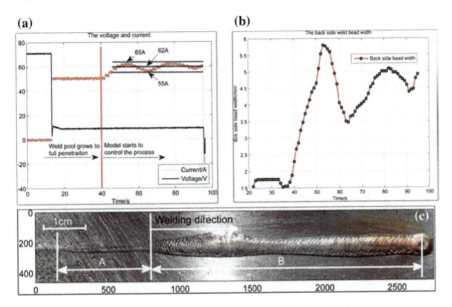

Fig. 9 Electric parameters, backside of weld pool and weld bead shape [58]

characteristic parameters of weld pool was identified. Furthermore, the adjusted rule was transferred to the control algorithm performed by robot. The variation of weld pool and weld quality was controlled by welding robot mimic a human welder [57–59]. The results are shown in Fig. 9.

Although the linear model can reflect the adjustment of human welder on the weld pool to some extent and effective control results were obtained, as we know, the adjusted process of human welder on the weld pool reflects nonlinear characteristics and is a multi-parameter coupled process. Therefore, the linear model is not able to describe the nonlinear features of human welder adjustment. Observations of manual welding process show that the welding current keeps constant and the weld torch attitude is more adjusted. Liu et al. [60, 61] elaborated the principle of intelligent welding and modeling of human welder's responses and set up a MIMO control system in which welding current, arc length, and welding speed set as the output and the weld pool parameters set as the input of system. The weld pool, at the external disturbance, was controlled lastly by a fuzzy-neuro networks controller and a continuous penetration weld bead was obtained, demonstrating the effectiveness of the nonlinear model. The control results are shown in Fig. 10.

Zhang and Liu mainly regulated the heat input of weld pool and cannot elaborate the mechanism of changing of weld pool status adequately by the linear or nonlinear neuro networks model. In their future works [62–64], the model referenced adaptive control algorithm, adaptive fuzzy-neuron networks and local iterative adaptive fuzzy neural network was adapted to model the human welder's responses and control the

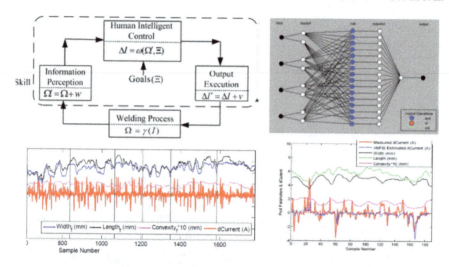

Fig. 10 Control results of weld pool by fuzzy-neuron networks [61]

weld quality, respectively. Virtual welding method was also utilized to model and learn the human welder behavior for realizing the intelligent robot welding [65].

It is known that the dynamic variation of weld pool concurrently includes the thermal and force change as well as the coupled change of thermal and force. The welding torch orientation is an appropriate physical variable for characterizing the thermal-force coupled change because of the change of heat input distribution and force direction. Thereby, how to measure the change of weld torch orientation becomes very urgent to intelligentize the robot welding. For this purpose, with the limited search for articles, some investigations have been reported. Byrd et al. [66] developed a virtual simulation system and trained the novel welder as a skilled human welder in a short period. Hashimoto [67] utilized a six-axis curve sensor and electromyography sensor to sample the welding torch motion of the skilled human welder and established a database using those sampled data. However, building this database needs to prepare lots of data of the skilled human welders, the task is very hugeness. Furthermore, Hahsimoto [68] adapted four cameras to real-time capture the traveling trace of weld torch in welding process and identified the difference of skilled or unskilled human welder via compared the torch orientation trace. Seto et al. [69] qualitatively analyzed the responses of human welder in welding process by means of analytic hierarchy process (AHP). Above investigations only provided a method for measuring the welding torch attitude and focused on the human welder skill training. The physical model between welding torch orientation and weld quality has still not been established. In recent reports, University of Kentucky Zhang et al. [70] further studied the measurement of the welding torch orientation and built a machine–human interactive operating system for allowing the welding robot to mimic the human welder operation and obtain the corresponding data of weld pool characteristics, welding speed, welding current and welding torch attitude. Based

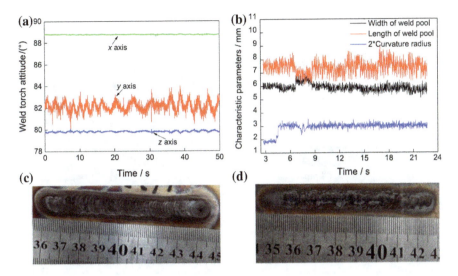

Fig. 11 Experimental results with skilled welder. **a** Weld torch attitude; **b** weld pool surface characteristic parameters; **c** topside weld appearance; **d** backside weld appearance [73]

on the acquired data, a data-driven nonlinear model of human welder's responses was first established to analyze the human welder's behavior, and the effectiveness of model was tested by the varying current and arc length experiments [71, 72]. Although the human welder's response was nonlinearly modeled in above literatures and its effectiveness was demonstrated by experiments, the mechanism of welder adjustment was not clear, which mainly affects the control of weld pool status and the achievement of high-quality welds.

To clearly reveal the mechanism of human welder's adjustment on the weld pool and change the weld pool status and keep the weld quality, LUT Zhang et al. [73] first established a synchronously measurement system to obtain the welding torch attitude and the 3D weld pool surface using a wire inertial measuring unit (IMU) and a laser vision-based approach, and then welding skills and responses of human welder were carefully analyzed from the changes of welding torch orientation and weld pool characteristics parameters. Analysis indicated that the change of torch attitude represents the welder's operating skills, welding experience, and smart decision. The three characteristic parameters reflect the welder's reactive response on the torch adjustments. The curvature radius of laser stripes can predict the changing trends of the weld pool surface, providing the needed information for welders to make a smart decision. The proposed scheme is feasible for measuring and analyzing the welder's skills and experience. The result is shown in Fig. 11.

Zhang [74] further simulated the internal essence of the change of welding torch orientation on the variation of thermal-force coupled system from the temperature field and force field. The results are shown in Fig. 12.

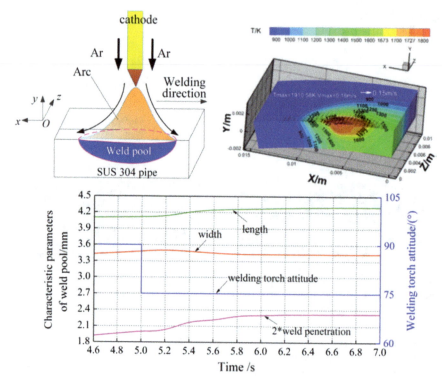

Fig. 12 Simulation of influence of weld torch orientation on weld pool [74]

On the other hand, the period of weld pool changed from a stimulus state to equilibrium state can be calculated and used to optimize the control algorithm, shorten control time for realizing the fast adjustment of welding robot like a human welder.

On the physical model simulation basis, the adjustment and responses of human welder were analyzed by frequency spectrum. The results show that the high-frequency adjustment of human welder cannot be responded by the weld pool, and the weld pool system can be considered as an inertial system. Hence, the sampling algorithm of welding torch attitude may be simplified, and the data is optimized. An adaptive kernel radius basis function neural networks model of welding torch orientation, welding speed and characteristic parameters of weld pool was established, and its effectiveness and accuracy were demonstrated by the varying welding current and welding speed experiments.

4 Conclusion

Sensing of weld pool benefits to automatic control of welding process; especially, real-time measurement of the weld metal fluid flow is a key factor to effectively control of weld quality and intelligentize robot welding, which needs to be urgently sought and developed.

The essence of his/her behavior represents a dynamic interactive evolution between human welder and weld pool, which can be numerically illustrated. Hence, a coupled mathematical model of human welder responses needs to be established in future investigations.

The nonobjective human welder's experience is fully extracted and transferred to a controlling strategy or algorithm which can be roundly performed by a welding robot like a human welder. It provides a new method to intelligentize robot welding.

Intelligent multi-sensor information fusion may realize the collection and analysis of many isolated information and then form the optimal determination for allowing welding robot to adapt more complicated conditions.

In robotic control system, the traditional control approaches, BP neural network is still utilized. Deep learning method will be applied and developed in intelligent welding manufacturing in the future.

To realize the optimization of welding robot system in time, space, information and function and improve the working efficiency as well as product quality of the welding robot, developing the multi-intelligent welding robot system is necessary and urgent.

Acknowledgements This work is funded by the National Natural Science Foundation of China (61365011 and 51775256) the Scientific Research Project of Colleges and Universities of Gansu Province (2018A-018), and Hongliu Outstanding Young Talents Support Project of Lanzhou University of Technology.

References

1. Chen SB, Zhang Y, Lin T et al (2004) Welding robotic systems with visual sensing and real-time control of dynamic weld pool during pulsed GTAW. Int J Robot Autom 19(1):28–32
2. Chen SB, Zhang Y, Qiu T et al (2003) Robotic welding systems with vision-sensing and self-learning neuron control of arc welding dynamic process. J Intell Rob Syst 36(2):191–208
3. Chang D, Son D, Lee J et al (2012) A new seam-tracking algorithm through characteristic-point detection for a portable welding robot. Robot Comput Integr Manuf 28(1):1–13
4. Moon SB, Hwang S, Shon W et al (2003) Portable robotic system for steel H-beam welding. Ind Robot 30(3):258–264
5. Pietrzak KA, Packer SM (1994) Vision-based weld pool width control. ASME J Eng Ind 116(2):86–92
6. Reisgen U, Purrio M, Buchholz G (2014) Machine vision system for online weld pool observation of gas metal arc welding processes. Weld World 58(5):707–711
7. Hong Y, Chang B, Peng G et al (2018) In-process monitoring of lack of fusion in ultra-thin sheets edge welding using machine vision. Sensors 18(8):1–18

8. Ju JB, Suga Y (2002) Penetration control by monitoring molten pool oscillation in TIG arc welding. In: Proceedings of the twelfth (2002) international offshore and polar engineering conference Kitakyushu, pp 26–31
9. Ju JB, Hiroyuki H, Suga Y (2004) Oscillation of molten pool by pulsed assist gas oscillating method and penetration control using peculiar frequency. J High Temp Soc 30(5):263–269
10. Alfaro SCA, Franco FD (2010) Exploring infrared sensing for real time welding defects monitoring in GTAW. Sensors 10(6):5962–5974
11. Kumar A, Sundarrajan S (2009) Optimization of pulsed TIG welding process parameters on mechanical properties of AA 5456 aluminum alloy welds. Mater Des 30:1288–1297
12. Fan CJ, Lv FL, Chen SB (2009) Visual sensing and penetration control in aluminum alloy pulsed GTA welding. Int J Adv Manuf Technol 42:126–137
13. Shen HY, Ma HB, Lin T et al (2007) Research on weld pool control of welding robot with computer vision. Ind Robot 34(6):467–475
14. Du QY, Chen SB, Lin T (2006) Inspection of weld shape based on the shape from shading. Int J Adv Manuf Technol 27(7–8):667–671
15. Li LP, Chen SB, Lin T (2005) The modeling of welding pool surface reflectance of aluminum alloy pulse GTAW. Mater Sci Eng A 394:320–326
16. Li LP, Chen SB, Lin T (2005) The light intensity analysis of passive visual sensing system in GTAW. Int J Adv Manuf Technol 27:106–111
17. Wang JJ, Lin T, Chen SB (2005) Obtaining of weld pool vision information during aluminum alloy TIG welding. Int J Adv Manuf Technol 26:219–227
18. Zhao DB, Yi JQ, Chen SB (2003) Extraction for three-dimension parameters for weld pool surface in pulsed gtaw with wire filler. J Manuf Sci Eng 125(3):493–503
19. Zhao DB, Chen SB, Wu L et al (2001) Intelligent control for the double-sided shape of the weld pool in pulsed gtaw with wire filler. Weld J 80(11):253s–260s
20. Chen SB, Lv N (2014) Research evolution on intelligentized technologies for arc welding process. J Manuf Process 16:109–122
21. Lin T, Chen HB, Li WH et al (2009) Intelligent methodology for sensing, modeling, and control of weld penetration in robotic welding system. Ind Robot 36(6):585–593
22. Kong M, Chen S (2009) Al alloy weld pool control of welding robot with passive vision. Sensor Rev 29(1):28–37
23. Richardson RW, Gutow DA (1984) Coaxial arc viewing for process monitor in gun control. Weld J 63(3):43–50
24. Gao J (2000) Vision-based measurement of weld pool geometry in TIG welding. Acta Metall Sin 12:18–24
25. Gött G, Uhrlandt D, Kozakov R et al (2013) Spectral diagnostics of a pulsed gas metal arc welding process. Weld World 57(2):215–221
26. Luo H, Lawrence F, Mun K et al (2000) Vision based GTA weld pool sensing and control using neuro-fuzzy logic. In: SIMTech technical report
27. Li LP, Lin T, Chen SB et al (2006) Surface height acquisition of welding pool based on shape from shielding (SFS). J Shanghai JiaoTong Univ 40(6):97–107
28. Li LP (2007) The method research on the pulse GTAW weld pool shape recover of 2A14 aluminum alloy from the image. Modern Weld 4:76–78
29. Mnich C, Al-Bayat F, Debrunner C et al (2004) In-situ weld pool measurement using Stereovision. In: Proceedings of the Japan–USA Symposium on Flexible Automation, ASME, Denver, Colorado, pp 19–21
30. Zhang YM, Li L, Kovacebic R (1995) Monitoring of weld pool appearance for penetration control. In: 4th international conference on trends in welding research, pp 5–8
31. Kovacebic R, Zhang YM (1997) Real-time image processing for monitoring of free weld pool surface. J Manuf Sci Eng 119(5):161–169
32. Zhang YM, Beardsley H (1994) Real-time image processing for 3D measurement of weld pool surface. Manuf Sci Eng 68(1):255–262
33. Saeed G, Lou MJ, Zhang YM (2004) Computation of 3D weld pool surface from the slope field and point tracking of laser beams. Meas Sci Technol 15(2):389–403

34. Song HS, Zhang YM (2008) Measurement and analysis of three-dimensional specular gas tungsten arc weld pool surface. Weld J 87(4):85–95
35. Zhang WJ, Liu YK, Zhang YM (2012) Characterization of three-dimensional weld pool surface in gas tungsten arc welding. Weld J 91:195s–203s
36. Zhang G, Shi Y, Huang JK et al (2014) Method to measure three-dimensional specular surface of TIG weld pool based on dot matrix laser pattern. J Mech Eng 50(24):10–14
37. Zhang G, Shi Y, Li CK et al (2014) Research on the correlation between the status of three dimensional weld pool surface and weld penetration in TIG welding. Acta Metall Sin 50(8):995–1002
38. Li DJ, Lu SP, Li DZ et al (2014) Principles giving high penetration under the double shielded TIG process. J Mater Sci Technol 30:172–178
39. Naito Y, Mizitani M, Katayama SJ (2003) Observation of keyhole behavior and melt flows during laser-arc welding. In: Proceedings of ICALEO, pp 1–9
40. Tsai CL, Hou CA (1998) Theoretical analysis of weld pool behavior in the pulsed current GTAW process. J Heat Trans-T ASME 110:160–165
41. Aval HJ, Farzadi A, Serajzadeh S et al (2009) Theoretical and experimental study of microstructures and weld pool geometry during GTAW of 304 stainless steel. Int J Adv Manuf Tech 42:1043–1051
42. Nguyen TC, Weckman DC, Johnson DA et al (2006) High speed fusion weld bead defects. Sci Technol Weld Join 11:618–633
43. Wu CS, Chen J, Zhang YM (2007) Numerical analysis of both front and back-side deformation of fully-penetrated GTAW weld pool surfaces. Comp Mater Sci 39:635–642
44. Traidia A, Roger F, Schroeder J et al (2013) On the effects of gravity and sulfur content on the weld shape in horizontal narrow gap GTAW of stainless steels. J Mater Process Tech 213:1128–1138
45. Xu G, Hu J, Tsai HL (2012) Modeling three-dimensional plasma arc in gas tungsten arc welding. J Manuf Sci Eng 134:031001–031013
46. Abed-AlKareem SS, Hussain IY (2007) Evaluation of temperature distribution and fluid flow in fusion welding processes. Int J Eng Sci 13:1313–1329
47. Zhang RH, Fan D (2007) Numerical simulation of effects of activating flux on flow patterns and weld penetration in ATIG welding. Sci Technol Weld Join 12:15–23
48. Yin XQ, Gou JJ, Zhang JX et al (2012) Numerical study of arc plasmas and weld pools for GTAW with applied axial magnetic fields. J Phys D Appl Phys 45:285203
49. Liu LY (2012) Application study of real time dynamic control of weld pool using laser vision method. J Hunan Univ Sci Eng 12:26–28
50. Yue JF, Zhang CX, Li LY (2009) MIG welding robot welding quality control based on multi-–information fusion. China Mech Eng 9:1112–1115
51. Zhang Y, Chen SB, Qiu T et al (2002) Study on real-time control of welding pool in welding flexible manufacturing cell. Trans China Weld Inst 4:1–5
52. Shi Y (2007) Research on the intelligent welding control system of Aluminum pulsed MIG welding. Dissertation, LanZhou University of Technology
53. Du QY (2006) Extraction and intelligent control of 3D dynamic weld pool shape information of pulsed GTAW with wire filler. Dissertation, Shanghai Jiao Tong University
54. Fan D, Shi Y, Lu LH et al (2010) Multi-information fusion and decoupling control of pulsed MIG welding process. Weld Join 5:8–12
55. Lu L, Fan D, Huang J et al (2012) Decoupling control scheme for pulsed GMAW process of aluminum. J Mater Process Tech 212(4):801–807
56. Zhang WJ, Zhang YM (2012) Modeling of human welder response to 3D weld pool surface: part 1-principles. Weld J 91:310s–318s
57. Zhang WJ, Zhang YM (2012) Modeling of human welder response to 3D weld pool surface: part 2-results and analysis. Weld J 91:329s–337s
58. Zhang WJ, Zhang YM (2013) Dynamic control of GTAW process using human welder response model. Weld J 92:154s–166s

59. Liu YK, Zhang YM (2014) Skilled human welder intelligence modeling and control: part 1-modeling. Weld J 93:46-s-52-s
60. Liu YK, Zhang YM (2014) Skilled human welder intelligence modeling and control: part 2-analysis and control application. Weld J 93:162s–170s
61. Liu YK, Zhang YM (2014) Model-based predictive control of weld penetration in gas tungsten arc welding. IEEE Trans Control Syst Technol 22:955–966
62. Liu YK, Zhang WJ, Zhang YM (2015) Dynamic neuro-fuzzy based human intelligence modeling and control in GTAW. IEEE Trans Autom Sci Eng 12:324–335
63. Liu YK, Zhang YM (2013) Control of human arm movement in machine-human cooperative welding process. Control Eng Pract 21:1469–1480
64. Liu YK, Zhang WJ, Zhang YM (2013) Adaptive neuro-fuzzy inference system (ANFIS) modeling of human welder's responses to 3D weld pool surface in GTAW. J Manuf Sci Eng Trans ASME 135:0210101–02101011
65. Liu YK, Shao Z, Zhang YM (2014) Learning human welder movement in pipe GTAW: a virtualized welding approach. Weld J 93:388s–398s
66. Byrd AP, Stone RT, Anderson et al (2015) The use of virtual welding simulators to evaluate experienced welders. Weld J 12:389s–395s
67. Hashimoto N (2015) Measurement of welder's movement for welding skill analysis. Bull Hiroshiima Inst Tech Res 49:83–87
68. Hashimoto N (2015) Difference of improving welder's skill through training progression. Bull Hiroshima Inst Tech Res 49:75–81
69. Seto N, Mori K, Hirose S (2012) Extracting of the skill from expert welders for aluminium alloy and investigation of their viewpoints using analytic hierarchy process. Light Metal Weld 8:14–22
70. Zhang WJ, Xiao J, Chen HP et al (2014) Measurement of three-dimensional welding torch orientation for manual arc welding process. Meas Sci Technol 3:031510–031527
71. Liu YK, Zhang YM (2015) Iterative local ANFIS-based human welder intelligence modeling and control in pipe GTAW process: a data-driven approach. IEEE Trans Mech 20:1079–1088
72. Liu YK, Zhang YM (2015) Controlling 3D weld pool surface by adjusting welding speed. Weld J 94:125s–134s
73. Zhang G (2017) Human welder's response-based interactive behavior analysis of welder with weld pool and control of weld pool morphology. Dissertation, Lanzhou University of Technology
74. Zhang G, Shi Y, Gu YF et al (2017) Welding torch attitude-based study of human welder interactive behavior with weld pool in GTAW. Robot Comput Integr Manuf 48:145–156

Prof. Ding Fan Ph.D tutor, works at Lanzhou University of Technology, who is the member of IIW-IX-H and IIW-SG212, and the Executive director of China welding society, deputy director of welding robot and automation specialty committee, member of expert committee of national welding industry alliance, director of welding specialty committee of Gansu province. Osaka University for further study and cooperative research for three times and served as a guest researcher. He focuses on the following studies. Welding method and welding physics, intelligent control and automatic robot welding, laser-based material processing technology. He has published more than 300 papers which were accepted by Journal of Physics D, Welding Journal, Heat and Mass Transfer and so on. There are 10 China patterns applied and authorized.

Research Papers

Spectral Analysis of the Plasma Emission During Laser Welding of Galvanized Steel with Fiber Laser

Bo Chen, Zhiwei Chen, Han Cheng, Caiwang Tan and Jicai Feng

Abstract Spectral information during the laser welding of galvanized steel was obtained. The plasma spectra were analyzed under different laser power and sheet gaps; signal filter method and statistical method were used to process the obtained spectral information. It was found that the plasma temperature increased with the increment of laser power, and plasma temperature and spectral intensity would be the least under proper sheet gaps. Statistical process control method was used to analyze the relationship between the welding quality and the spectral information, and it was found that the welding defects of the laser welding of galvanized steel could be automatically detected by the spectral information.

Keywords Laser welding · Galvanized steel · Spectral analysis · SPC

1 Introduction

Laser beam welding of metal sheets has been widely used in the automobile manufacturing industry. Galvanized steel has become one of the most widely used materials in the automobile industry because of its low cost, availability, and corrosion resistive properties, and the utilization rate has reached more than 60%. However, during the laser welding of galvanized steel, corrosion-resisting zinc coatings will boil, and if zinc vapor is trapped in the faying surface region during lap welding, it will vent through the weld pool and cause gross porosity or even periodically ejecting the liquid steel [1]. Nowadays, the welding quality was mainly done by manual inspection and relied on destructive testing which is time-consuming. Online monitoring will be

---

B. Chen (✉) · C. Tan · J. Feng
State Key Laboratory of Advanced Welding and Joining, Harbin Institute of Technology, Harbin 150001, China
e-mail: chenber@163.com

B. Chen · Z. Chen · H. Cheng · C. Tan · J. Feng
Shandong Provincial Key Laboratory of Special Welding Technology, Harbin Institute of Technology, Weihai 264209, China

© Springer Nature Singapore Pte Ltd. 2019
S. Chen et al. (eds.), *Transactions on Intelligent Welding Manufacturing*, Transactions on Intelligent Welding Manufacturing, https://doi.org/10.1007/978-981-13-7418-0_3

helpful to ensure the quality of the laser welding process, and various on-line monitoring methods have been proposed, such as infrared camera [2–4], electrical and capacitive sensors [5], acoustic monitoring of waves generated by keyhole plasma [6, 7], CCD detection of the seam [8, 9], ultrasounds sensor [10–12], etc. Gu et al. [13] used fast Fourier transform to analyze the acoustic signal generated during the laser welding of Al 1100 alloy and found that the range of 3–9 kHz was associated with the closure of the keyhole. Liu [14] developed a mathematical model to correlate AE with martensitic transformation, and the amplitude of the AE signal was found to be related to the distance between the position of the transformation and the sensor location. Mori et al. [15] used five photodiodes to detect the welding defects generated during the CO_2 laser welding process, and defects such as underfill and pits could be detected by the method. Park et al. [16, 17] detected the plasma generated in CO_2 laser welding by using ultraviolet and infrared photodiodes, and they classified the welding qualities into five different categories.

Nowadays, the quality of welding was assured by off-line destructive and nondestructive evaluation techniques, like penetrant liquids, X-rays, ultrasonic, magnetic particles, or macrographs [18]. The research for an efficient on-line welding quality monitoring system able to detect the defects early during the welding process has become an active research area. The plasma plume generated during the laser welding process contains lots of information that could reflect the welding quality. The relationship between the welding quality and the spectral information has not been well investigated to date. Automatic control of the laser welding of galvanized steel has not been achieved. Spectrometer has been proved to be an effective way of monitoring the CO_2 laser welding process [19], while few studies have been conducted with Nd:YAG or fiber laser. Sibillano et al. [20] studied the relationship between electron temperature and penetration depth by using spectroscopic analysis. Lober et al. [21] used spectrometer to study the energy transfer of the plasma induced by a 10 kW CO_2 laser, and the absorption and the refraction of the laser beam were determined. Ancona et al. [22] presented a study for CO_2 laser welding defects detection using a spectrometer, and the electron temperature was calculated to correlate with the welding defects such as lack of penetration, welding disruptions, and seam oxidation.

This paper studied the spectra generated during the fiber laser welding of galvanized steel, the influences of laser power and sheet gaps on the spectral intensity, and electron temperature were studied, and a novel method based on statistical process control (SPC) theory was used to detect the defects generated during the welding process automatically.

2 Experimental Details

The galvanized steel used in this paper was DX51D+Z, the metals were cut to small plates of 150 mm × 50 mm × 0.8 mm, and the chemical composition of the steel was shown in Table 1.

Table 1 Chemical composition and mechanical property of galvanized steel [23]

Type	Chemical composition (%)						Mechanical properties		
	C	Si	Mn	P	S	Ti	Tensile strength (Mpa)	Yield strength (Mpa)	Elongation (%)
DX51D+Z	0.12	0.5	0.6	0.1	0.045	0.3	273	353	33

A 6 kW fiber laser unit with a laser beam diameter of 0.2 mm was used to do the laser welding experiment. To prevent oxidation, Argon gas was used as the shielding gas. KUKA KR-60HA robot was used as the moving mechanism. The Avantes spectrometer was used to obtain the spectrum during the laser welding process. The resolution ratio of the spectrometer was 0.1 nm, and the detectable spectral range of the spectrometer was 310–425 nm, the integration time was 2 ms.

During the laser welding process, laser power was the most important factor that influenced the welding quality and the spectral intensity. The influence of the laser power was studied in the paper to investigate the relationship between laser power and spectral information. Overlap welding was done along the length direction to obtain the spectral information during different welding conditions.

Gaps between the plates were important factors that influence the welding quality of galvanized steel welding. Spacers of different thickness were used to obtain different gaps. The spacers between 0.1 mm thick and 0.5 mm thick were used to generate various gap conditions between overlap joints. The spacers were inserted at both ends along with longitudinal direction. During the experiment, the process parameters such as shielding gas flow rate and welding speed were fixed. The spectral signals in each experiment were recorded under the fixed sampling time described above.

3 Experimental Results and Discussion

3.1 Spectra Measurements

To analyze the spectrum of laser welding process, the spectra should be collected exactly. Although the spectrometer has high accuracy, because of various conditions such as atomic physics and spectral distortion, the measured spectrum line is different from the standard spectrum. Some conditions are unavoidable, to improve the measurement accuracy, the spectrum should be carefully selected. The spectral lines used in this paper are listed in Table 2.

Table 2 Characteristic parameters of Zn III spectral line

Wavelength λ (nm)	Transition probability A (s^{-1})	Upper statistic weight g	Upper level E (eV)
358.1	1.12×10^9	9	32.55
363.0	2.91×10^7	5	32.47
372.0	1.59×10^9	9	32.55
374.9	5.57×10^8	5	32.86
382.4	4.80×10^8	5	32.36
386.0	2.34×10^8	3	32.76
390.0	2.50×10^6	7	35.16

3.2 Electron Temperature Calculation by Boltzmann Method

The plasma temperature is commonly used to characterize the laser-induced plasma during additive manufacturing, which can be calculated by Boltzmann double-line method when the plasma is in the local thermodynamic equilibrium (*LTE*) assumption state [19]. The Boltzmann double-line method is expressed as follows:

$$\frac{I_1}{I_2} = \frac{A_1 g_1 \lambda_2}{A_2 g_1 \lambda_1} \exp\left(-\frac{E_1 - E_2}{kT}\right) \tag{1}$$

where I_1, I_2 are radiation intensity; A_1, A_2 are transition probability; g_1, g_2 are degeneracy of the excited levels; λ_1, λ_2 are wavelengths; E_1, E_2 are excitation energy; and k is the Boltzmann constant. However, there are some certain errors in calculation result of this approach. The Boltzmann plot method was selected for improving accuracy instead of Boltzmann double-line method by using the relative intensities of several emission lines rather than only two lines [20]. It could be described as the following equation [21]:

$$\ln\left(\frac{I_{mn} \lambda_{mn}}{g_m A_{mn}}\right) \propto -\frac{E_m}{kT} + \ln\left(\frac{Nhc}{z(T)}\right) \tag{2}$$

where m is excited level; n is lower state level; λ_{mn} is wavelength; A_{mn} is transition probability; g_m is degeneracy of the excited level; E_m is upper state energy level; T is plasma temperature; N is the number density of the species involved; h is Planck's constant; c is the speed of light; and $z(T)$ is the partition function of the element at temperature T. So the plasma temperature can be calculated from the slope of the linear function of Eq. (2).

In this paper, in order to calculate the plasma temperature more accurately, six neutral Fe I emission lines were chosen according to the Boltzmann plot method. They were in the range of 375–390 nm, which are within 15 nm so that they are close enough to be considered as a linear correction on the intensities and have

Table 3 Spectroscopic data of neutral Fe I emission lines

λ (nm)	A_{mn} (s^{-1})	E_n (eV)	E_m (eV)	g_m
375.694	2.66E+08	3.57	6.87	11
383.926	2.55E+08	3.05	6.28	9
385.921	1.11E+08	2.40	5.62	11
387.250	5.25E+07	0.99	4.19	5
387.802	5.41E+07	0.96	4.15	7
388.628	3.71E+07	0.05	3.24	7

considerable differences of the excitation energy level to minimize the fitting error in the Boltzmann plot. And all of them were summarized in Fig. 1 and Table 3 [22].

Boltzmann plot of a randomly selected sampling point, when 700 W laser power, 7.78 g/min powder feeding rate, and 5 mm/s traverse speed were used in single-layer, single-trace experiment, is shown in Fig. 2. The plasma temperature computed was 13,764.9 K, as the slope of the linear fit included in the figure is −0.8423. However, the calculated temperature by the Boltzmann double-line method was 13,957.6 K. In this paper, the Boltzmann plot method was mainly used for the research on the parameter influence laws. So, the plasma temperatures under different technical conditions could be similarly acquired.

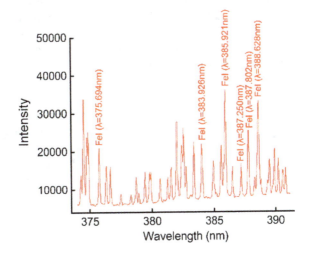

Fig. 1 Spectroscopic parameters of the Fe I emission lines used to calculate the plasma temperature

3.3 Influence of Laser Power to the Quality of Laser Welding of Galvanized Steel

To explore the influence of the laser power on the spectral intensity, different laser powers were used to do the welding experiments. The welding speed was 2.5 m/s, the argon flow rate was kept 20 L/min, and the overlap welding experiment was done to obtain the spectra under different experimental conditions. Table 4 shows the relationship between different laser power and the corresponding spectral intensity obtained by mean value of the seven Zn spectra in Table 2.

It could be seen that the welding quality could be divided into three classes: incomplete penetration, good quality, and burn-through. When the laser power was between 3200 and 3600 W, because the laser power was too low and the heat cannot be achieved to obtain good welding quality, defects of incomplete penetration occurred. With the increment of laser power, the spectral intensity increased. When the laser power was between 3700 and 4500 W, the laser power was appropriate and good welding forming quality was obtained, however, because the zinc vapor escaped, spatters occurred and some pinholes generated; when the laser power was between 4600 and 4900 W, because the laser power was too high, defects of burn-through occurred. Figure 3 shows the welding appearance under different laser power.

Figure 4 shows the obtained spectra under different laser power, and Fig. 5 shows the relationship between different laser power and spectral intensity under different conditions. It could be seen that the spectral intensity increased with the increment of laser power under incomplete welding and normal conditions; however, when the burn-through defects occurred, the intensity decreased with the increment of laser power, that is because the zinc evaporated faster with more heat input, and that leads to the decrease of spectral intensity.

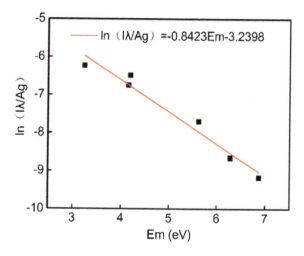

Fig. 2 Boltzmann plot using six neutral iron lines from Table 4

Figure 6 shows the calculated temperature under different welding conditions of different laser power, the electron temperature increased with the increment of the laser power under different penetration status. During the normal penetration status, the electron temperature almost kept constant. However, the temperature of the burn-through conditions was much lower than the incomplete welding condition and the normal condition.

Table 4 Relationship between laser power and spectral intensity

No.	Power (kW)	Relative spectral intensity	Weld appearance
1	3280	152,674	Not welded
2	3400	159,655	Incomplete penetration
3	3520	168,380	Incomplete penetration
4	3640	223,541	Incomplete penetration
5	3880	155,561	Normal
6	4000	176,304	Normal
7	4360	203,180	Normal
8	4480	212,901	Little spatter
9	4600	242,599	Spatter and pinhole
10	4720	191,932	Burn-through
11	4860	25,239	Burn-through
12	4960	21,772	Burn-through

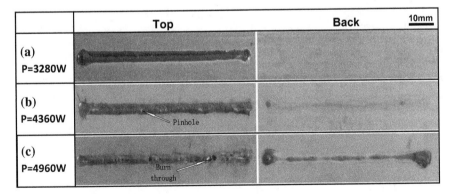

Fig. 3 Weld appearance quality under different laser power

3.4 Influence of Plates Spaces to the Spectra

The spaces between plates have an important influence on the forming quality during laser welding of galvanized steel. In lap welding of galvanized steel, the zinc layer evaporates violently at the weld interface and results in undesirable high porosity in the welded joints. The gap between the two steel plates allows the explosive zinc vapors to vent through the gaps. However, in commercial manufacturing environments, it is difficult to maintain a constant gap. The influence of the plate spaces on the spectral intensity and electron temperature was studied. The experiment was done with the laser power being kept constant at 4600 W, and welding speed kept 2.5 m/min, shielding Argon gas flow rate kept 20 L/min, and the experiment was done with plates spaces of 0.1, 0.2, 0.3, 0.4, and 0.5 mm. Table 5 shows the mean value of the obtained seven Zn spectra under different plates spaces.

Figure 7 shows the weld appearance of different plate spaces under the laser power of 4600 W; the spaces were 0, 0.2, 0.5 mm, respectively. Figure 8 shows the spectral intensity with 0, 0.1, 0.2, 0.3, 0.4, 0.5 mm spaces. Figure 9 shows the relationship between the intensity of Zn III spectra of 358.1 nm and different spaces. It could be seen that the relative intensity of Zn III spectra was, respectively, 23,654. 315, 182, 536, 647, 842 corresponding to the 0.1, 0.2, 0.3, 0.4 and 0.5 mm spaces. When there was no gap, the zinc vapor could not escape from spaces, and a lot of spatters

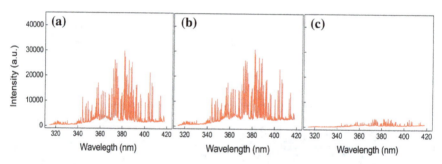

Fig. 4 Spectra intensity of different laser power: **a** $P = 3280$ W; **b** $P = 4360$ W; **c** $P = 4960$ W

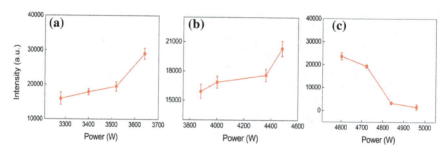

Fig. 5 Relationship between laser power and spectral intensity of 358.1 nm Zn III spectral line: **a** incomplete welding; **b** normal; **c** burn-through

Spectral Analysis of the Plasma Emission During Laser Welding ...

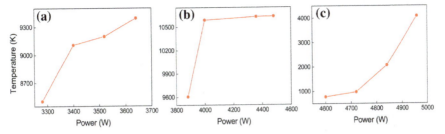

Fig. 6 Relationship between laser power and electron temperature: **a** incomplete welding; **b** normal; **c** burn-through

Table 5 Relationship between relative spectral intensity and plates spaces

No.	Spaces (mm)	Relative spectral intensity	Weld appearance
1	0.1	315	Not welded
2	0.2	182	Normal
3	0.3	536	Incomplete weld
4	0.4	647	Incomplete weld
5	0.5	842	Burn-through

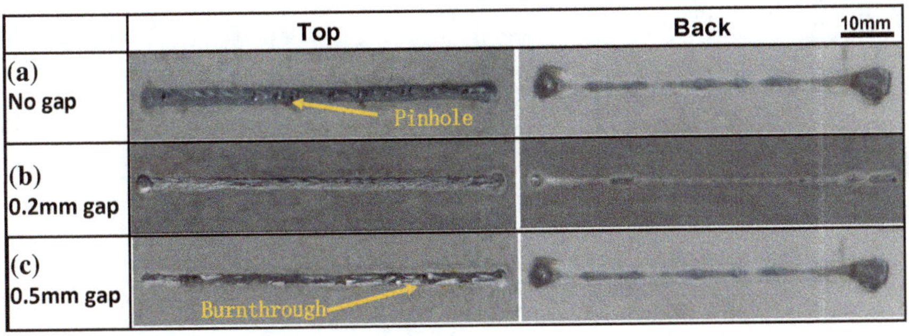

Fig. 7 Weld appearance of different plate spaces

occurred, that resulted in the higher intensity, and some pinholes occurred because the galvanized layer was lost, as shown in Fig. 7a. When the plate space was 0.2 mm, the zinc vapor could escape from the gap, almost no spatter occurred, the lowest spectrum intensity was reached because of the stable welding process, and the best welding quality was reached, as shown in Fig. 7b. When the plate space was 0.5 mm, because the plate space was too large, the zinc vapor evaporates immediately, the increased zinc plasma led to a higher spectral intensity; however, the higher zinc evaporation led to a buffer layer besides the weld pool, and the buffer layer hindered the fusion of the two plates, and burn-through defects occurred.

Figure 10 shows the relationship between the electron temperature and different plate spaces. When the space was 0.1 mm, the electron temperature was relatively

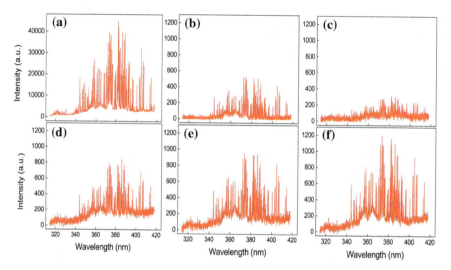

Fig. 8 Relative spectral intensity with different plate spaces: **a** 0 mm space; **b** 0.1 mm space; **c** 0.2 mm space; **d** 0.3 mm space; **e** 0.4 mm space; **f** 0.5 mm space

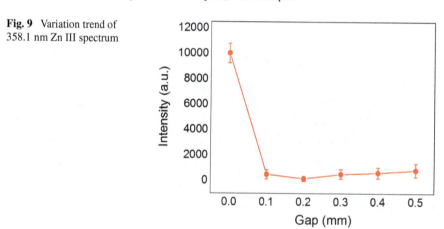

Fig. 9 Variation trend of 358.1 nm Zn III spectrum

high and dropped suddenly when the space became 0.2 mm; however, when the plate space was more than 0.2 mm, the electron temperature increases gradually with the spaces.

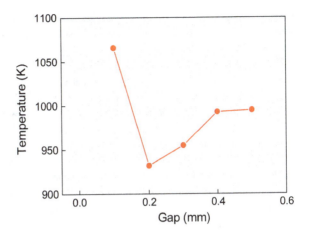

Fig. 10 Relationship of different spaces

3.5 Weld Quality Control Based on Statistical Process Control (SPC)

Statistical process control (SPC) is an industry standard methodology for measuring and controlling quality during the manufacturing process. Quality data in the form of process measurements is obtained in real time during manufacturing. This data is then plotted on a graph with pre-determined control limits. Data that falls within the control limits indicates that everything is operating as expected, and any variation within the control limits is likely to a common cause, the natural variation that is expected as part of the process.

(1) Spectrum information under different plate spaces

From the above analysis, it could be seen that when the sheet gaps were different, the relationship between the spectrum intensity and the gap was different, so the spectrum was analyzed under different sheet gaps.

Because the original spectrum contained a lot of noises, the spectrum was first filtered by a moving average filter to remove the noise in the signal, the equation form of the moving filter was shown in Eq. (3):

$$y[i] = \frac{1}{M} \sum_{j=-(M-1)/2}^{(M-1)/2} x[i+j] \qquad (3)$$

where $x[\ ]$ is the input signal, $y[\ ]$ is the output signal, and M is the number of points in the average. In this paper, M is equal to 9 because we found that it was the best value to obtain the optimal results. The filtered results were shown in Fig. 11.

Because there was a lot of disturbances during the start and end part of the welding process, the spectral information during the start and end part was removed from the curves to obtain the normal distribution of the spectral information with the sheet gaps 0, 0.2 and 0.5 mm, as shown in Fig. 12. Line 1 showed the normal distribution line of the whole data, and line 2 showed the distribution line of the data after being processed by subgroups.

The average spectral intensity was 16,292.54, and the variance was 1147.6 when the sheet gap was 0. So, the upper control line $+3\sigma$ was 19,735.35 and the lower control line -3σ was 12,849.73. The control line and the welding image were shown in Fig. 13. It could be seen that when the spectral intensity was within the upper and lower control lines, the welding quality was relatively good; and there are some points ran out of the control lines, when the spectral intensity exceeded the control lines, a lot of spatters occurred and some pinholes occurred at the corresponding locations. Among the 762 samples, 98 samples were below the lower control line and 43 samples exceeded the upper control line, so there was about 20% spatters occurred during the welding process.

Figure 14 shows the image and corresponding control chart when the sheet gap was 0.2 mm. It could be seen that the welding quality was relatively good, among the 1000 samples, only 7 samples were lower than the lower control line and 8 samples exceeded the upper control line.

Figure 15 shows the welding image and corresponding control charts when the sheet gap was 0.5 mm. The upper and lower control lines were 203.54 and 705.51, respectively. When the points crossed the control lines, burn-through defects occurred. Among the 1000 samples, 65 samples were lower than the lower control

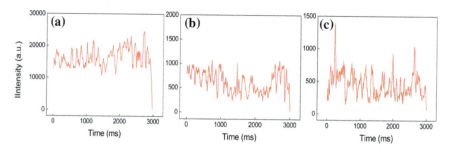

Fig. 11 Filtering waveform of different spaces: **a** no gap; **b** 0.2 mm gap; **c** 0.5 mm gap

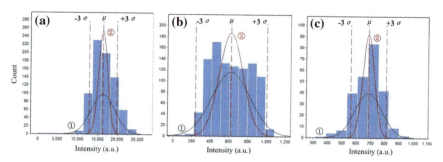

Fig. 12 Gaussian distribution

line and 32 samples exceeded the upper control line; there was about 10% burn-through defects occurred during the welding process.

From the results, it could be seen that when the plate gap was 0.2 mm, the spectral intensity was almost within the upper and lower control charts, and the welding process was almost stable. When the plate gap was 0 or 0.5 mm, the spectral intensity exceeded the control limits, which means defects such as spatter or burn-through occurred. Welding quality control could be realized by the spectral information.

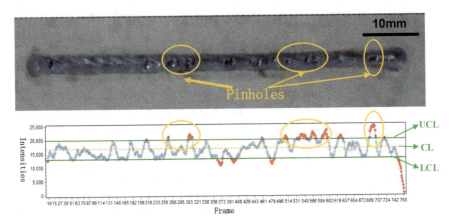

Fig. 13 SPC control $\bar{X}$–R_s result of Zn III spectrum of no gap

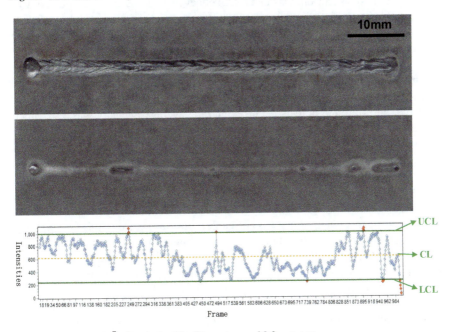

Fig. 14 SPC control $\bar{X}$–R_s result of Zn III spectrum of 0.2 mm gap

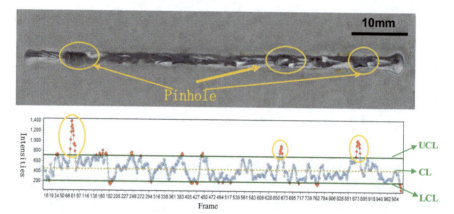

Fig. 15 SPC control $\bar{X}$ –R_s result of Zn III spectrum of 0.5 mm gap

(2) SPC control of gradually varied gap

The experiment was conducted under the following conditions: laser power 4600 W, defocusing distance +20 mm, weld speed 2.5 m/min, Argon flow rate 20 L/min, and spaces 0.1–0.5 mm.

Figure 16 shows the variation of 358.1 nm Zn III spectrum intensity with the time. It could be seen that when the time is from 0 to 1 s, corresponding to the gaps from 0.1 to 0.2 mm, the spectrum was in a relatively even area, the average intensity was 347, and the standard deviation was 284, the average intensity of the spectrum was relatively small, and the spectrum fluctuated in a range. That is because the space was appropriate and the zinc vapor could evaporate from the spaces, so the welding quality was relatively good. When the space was bigger than 0.2 mm, the intensity became very intense and fluctuated remarkably, spatters occurred and the welding quality was unstable. When the space was bigger than 0.5 mm, because the space was too large, burn-through defects occurred and the average intensity was 950, much bigger than the even area, and the standard deviation reduced to 62, a lot of zinc vapor evaporated from the space and the holes generated in the plates, that led to a bad welding quality.

From Fig. 16, it could be seen that there was a lot of noises in the spectrum, average filtering method was used to process the waveform. The spectrum after the filtering was shown in Fig. 17.

Because the spectral intensity was not stable during the start and end of the welding process, so 100 spectra of the start and end of the welding process were omitted to obtain the SPC control chart. Figure 18 shows the normal distribution of the 0–1 s spectral information after the smoothing. A rectangle in the figure represented the frequency of occurrence of a certain spectral intensity. It could be calculated from the figure that the average intensity μ was 685.23, the variance σ was 90.6, so the upper control limit $+3\sigma$ was 809.16, and the lower control limit -3σ was 561.92.

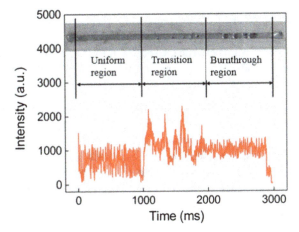

Fig. 16 Variation of 358.1 nm Zn III spectrum intensity with the gap changing

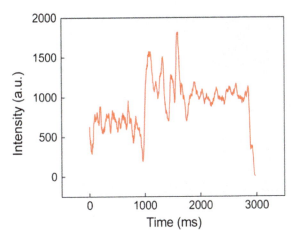

Fig. 17 Spectrum waveform after filtering

After the control chart was obtained, it could be used to detect the quality of the whole welding process, and the $\bar{X}$–R_s control chart was obtained as shown in Fig. 19. It could be seen that the spectral intensity fluctuated between the upper and lower control limits during the first one second, after that, the average spectral intensity exceeded the upper and lower control limits, and it could be inferred that the welding quality was unstable, and defects occurred, which was in accordance with the actual welding results.

4 Conclusion

Relationship between the spectral information and the welding quality of laser welding of galvanized steel was formed, and *SPC* method was applied to use the spectral information to detect the forming quality. Conclusions are drawn as follows:

(1) The spectral line intensity and the plasma temperature have relationship with the laser power and the welding quality. When the laser power was low and incomplete penetration occurred, in this case, the spectral line intensity increased with the laser power; while the laser power was too high and burn-through defects occurred, the spectral intensity decreased with the laser power. However, the plasma temperature increased with the increment of laser power regardless of the defects.

(2) Changes of the spectral intensity and the plasma temperature had relationship with the sheet gaps. The welding quality was the best when the sheet gap was 0.2 mm, while the spectral line intensity and the plasma temperature was the least.

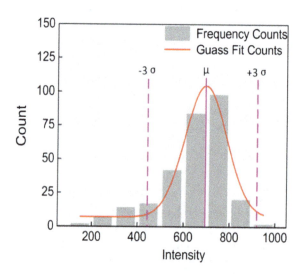

Fig. 18 Normal distribution of the spectral information

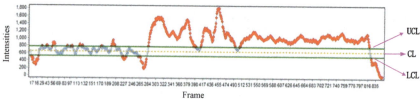

Fig. 19 $\bar{X}$ –R_s chart of Zn III spectrum

(3) There was a relationship between the processing parameters such as laser power, sheet gap and the spectral information; however, the relationship was segmented.
(4) The variance of the welding could be detected by the spectral information by using statistical process control method.

Acknowledgements This work was supported by the National Key R&D Program of China under the Grant (2018YFB1107900), the Shandong Provincial Natural Science Foundation, China (ZR2017MEE042), the Shandong Provincial Key Research and Development Program (2018GGX103026).

References

1. Li X, Lawson S, Zhou Y et al (2007) Novel technique for laser lap welding of zinc coated sheet steels. J Laser Appl 19(4):259–264
2. Bicknell A, Smith JS, Lucas J (1994) Infrared sensor for top face monitoring for weld pools. Meas Sci Technol 27(4):371
3. Chen Z, Gao X (2014) Detection of weld pool width using infrared imaging during high-power fiber laser welding of type 304 austenitic stainless steel. Int J Adv Manuf Technol 74(9–12):1247–1254
4. Chokkalingham S, Vasudevan M, Sudarsan S et al (2012) Predicting weld bead width and depth of penetration from infrared thermal image of weld pool using artificial neural network. Insight Non-Destr Test Cond Monit 54(5):272–277
5. Li L, Brookfield DJ, Steen WM (1996) Plasma charge sensor for in-process, non-contact monitoring of the laser welding process. Meas Sci Technol 7(4):615
6. Chen C, Kovacevic R, Jandgric D (2003) Wavelet transform analysis of acoustic emission in monitoring friction stir welding of 6061 aluminum. Int J Mach Tools Manuf 43(13):1383–1390
7. Gu H, Duley WW (1999) A statistical approach to acoustic monitoring of laser welding. J Phys D Appl Phys 29(3):556
8. Huang WQ, Zhang YS, Shao CJ (2001) Application of CCD vision sensor and image processing technology in seam automatic tracking. J Shenyang Polytech Univ 23(1):12–14
9. Bae KY, Lee TH, Ahn KC (2002) An optical sensing system for seam tracking and weld pool control in gas metal arc welding of steel pipe. J Mater Process Tech 120(1):458–465
10. Hand DP, Peters C, Jones JDC (1995) Nd:YAG laser welding process monitoring by non-intrusive optical detection in the fibre optic delivery system. Meas Sci Technol 6(9):1389
11. Ikeda T, Kojima T, Tu JF et al (2000) In-process monitoring of weld qualities using multi photo-sensor system in pulsed Nd:YAG laser welding
12. Naso D, Turchiano B, Pantaleo P (2005) A fuzzy-logic based optical sensor for online weld defect-detection. IEEE Trans Industr Inf 1(4):259–273
13. Gu H, Duley WW (1999) A statistical approach to acoustic monitoring of laser welding. J Phys D Appl Phys 29(3):556–560
14. Liu YN (1997) Experimental study of dual-beam laser welding of AISI 4140 steel. Weld J 76(9):342s–348s
15. Mori K, Sakamoto M, Miyamoto I (1994) In-process monitoring in laser welding of automotive parts. Proc SPIE 2306
16. Park H, Rhee S, Kim D (2001) A fuzzy pattern recognition based system for monitoring laser weld quality. Meas Sci Technol 12(8):1318
17. Park YW, Park H, Rhee S et al (2002) Real time estimation of CO_2 laser weld quality for automotive industry. Opt Laser Technol 34(2):135–142
18. Maldague XPV, Moore PO (2001) Nondestructive testing handbook: infrared and thermal testing

19. Sibillano T, Ancona A, Berardi V et al (2009) A real-time spectroscopic sensor for monitoring laser welding processes. Sensors 9(5):3376–3385
20. Sibillano T, Rizzi D, Ancona A et al (2012) Spectroscopic monitoring of penetration depth in CO_2 Nd:YAG and fiber laser welding processes. J Mater Process Technol 212(4):910–916
21. Lober R, Mazumder J (2007) Spectroscopic diagnostics of plasma during laser processing of aluminium. J Phys D Appl Phys 40(19):5917–5923
22. Ancona A, Spagnolo V, Lugarà PM et al (2001) Optical sensor for real-time monitoring of Co(2) laser welding process. Appl Opt 40(33):6019–6025
23. ASTM (1997) ASTM A366/A366M-97e1, standard specification for commercial steel (CS) sheet, carbon (0.15 maximum percent) cold-rolled

Bo Chen is an associate professor in Harbin Institute of Technology, Weihai. He received his B.S. degree in materials processing engineering from Shandong University, China, and received M.S. and Ph.D. degree in materials processing engineering from Shanghai Jiao Tong University, China. He once worked as a visiting scholar at the Center of Laser Aided Intelligent Manufacturing (CLAIM) in the University of Michigan, USA. His main research area is intelligentized materials processing technology, including weld automation and intelligentization, laser welding, and laser additive manufacturing.

A Method for Detecting Central Coordinates of Girth Welds Based on Inverse Compositional AAM in Tube-Tube Sheet Welding

Yu Ge, Yanling Xu, Huanwei Yu, Chao Chen and Shanben Chen

Abstract Aimed at the location and guidance problems of X-rays automatic inspection equipment in tube-tube sheet (TTS) welding, this paper presents a method for detecting central coordinates of girth welds based on inverse compositional AAM with a triaxial flaw detection device for tubular heat exchanger. The method includes the design of software framework, calibration algorithm, center detection algorithm, etc. Using the proposed center detection algorithm, the accuracy is verified with experimental data.

Keywords AAM · Inverse compositional algorithm · Center detection · Girth welds tube-tube sheet welding

1 Introduction

Tubular heat exchanger (condenser) is a heat exchange equipment widely used in power, petroleum, chemical, nuclear energy, and other fields. Quality of tube-to-tube sheet weld (TTS weld) is the key to ensuring long-term healthy operation of the entire device. Conventional welding method is to insert the tube into the hole of the tube sheet and then weld the tube with the tube sheet from the outside of the tube sheet. This method is often called end face welding between the tube and tube sheet, as shown in Fig. 1. Heat exchanger tubes are usually arranged compactly, and the diameter of it is usually small (10–40 mm). Under these circumstances, multilayer manual argon arc welding and TIG are mostly adopted. Because of the complexity and difficulty of welding process, it is easy to cause defects such as root incomplete fusion, slag inclusion, porosity, and crack [1]. These defects can easily lead to TTS

Y. Ge · Y. Xu (✉) · C. Chen · S. Chen
School of Materials Science and Engineering, Shanghai Jiao Tong University, Shanghai 200240, China
e-mail: ylxu@sjtu.edu.cn

H. Yu
Shaoxing Special Equipment Testing Institute, Shaoxing 312071, China

© Springer Nature Singapore Pte Ltd. 2019
S. Chen et al. (eds.), *Transactions on Intelligent Welding Manufacturing*, Transactions on Intelligent Welding Manufacturing,
https://doi.org/10.1007/978-981-13-7418-0_4

Fig. 1 TTS weld in tubular heat exchanger

weld failure in environments with long-term high temperature, high pressure, and corrosion.

In tubular heat exchanger, large amounts of TTS weld are arranged compactly, weld region is small, and weld structure is complicated. Although TTS weld quality has an important influence on the safe operation of heat exchangers, it is difficult to do nondestructive test to inspect its internal defect due to the special geometry of TTS weld. Conventional flaw detection methods include visual inspection and radiographic inspection in the form of spot check [2]. They expend low labor and time cost but reduce inspection efficiency obviously. Besides, research on automatic flaw detection system based on rays or ultrasound mostly focuses on identify defects through image processing [3–6]. There is little research focusing on guiding the detector to the center of TTS weld automatically to realize automation of the entire flaw detection process. Have a composite review on international research, existing radiographic inspection devices on TTS weld all have lower level of automation with lots of labor rework and low work efficiency. It has to remove the positioning clip of the detection probe and move it to the next tube manually after the inspection of every tube.

In the development of this TTS weld X-ray digital automatic flaw detection system, calibration of inspection device and camera was first accomplished. Then taking photographs of tubes by CCD camera and importing images into the software. AAM fitting algorithm based on inverse compositional algorithm will detect the center of tubes and output the coordinate data. After the above work, subsequent autonomous positioning and automatic guidance can carry on.

Calibration includes camera calibration and hand-eye calibration. Because guide device in this project is a triaxial device without rotation degrees of freedom, it is hard to adopt conventional camera calibration method to take photographs and get

3D information in various poses. In addition, when guiding the detection probe to insert it into the TTS weld, it only requires low accuracy of depth. So this paper fixes the distance between calibration board and CCD camera before camera calibration. The intrinsic matrix is calculated only by 2D information.

Active appearance models (AAM) was first put forward by Edwards in 1998. It is widely used in face recognition, expression recognition, and other deformable objects nowadays. Original AAM algorithm was not ideal in practical application which could not realize accurate estimation of constant linear relationship between error image and parameter increment. On the contrary, inverse compositional image alignment algorithm has got great feedback in feature recognition applications by reversing the roles of image intensity and base appearance model when computing the incremental warp.

2 Hardware and Software System of Flaw Detection Device

2.1 Hardware System

As shown in Fig. 2, the triaxial flaw detection device composes of well-known motion controller, Ray Man High Voltage Power Supply, Point Grey BFLY-PGE-12A2M-CS CCD camera, industrial computer, custom-designed triaxial motion device, etc.

The motion controller can control the stepping motor to realize that three axes move and stop at the same time. The high-voltage ETC supplies 700–3000 W 300 V–300 kV high voltage power to stepping motor. It should be noted that the self-designed triaxial motion device does not have rotational degree of freedom so that the calibration progress can be simplified (Fig. 3).

2.2 Software System

This paper integrates calibration and center detection in a software written in MAT-LAB so that users could perform calibration and acquire center coordinates offline conveniently. Software flow chart and software interface are shown as in Figs. 4 and 5. The software has two core modules: calibration and center detection. Calibration parameters and center coordinates are stored in.mat.

2.2.1 Calibration Module

Before using offline calibration module, calibration board photographs taken by CCD at different positions should be stored in PC. Parameters that should be set in software include import path, export path, file name, actual calibration board size, grid

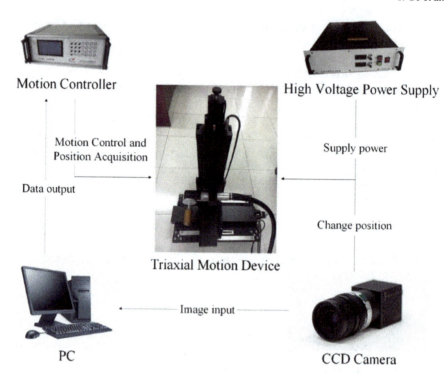

Fig. 2 Composition diagram of triaxial flaw detection system

Fig. 3 Installation position of CCD on triaxial device

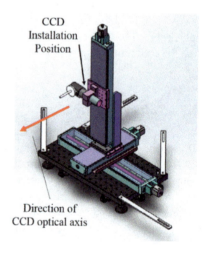

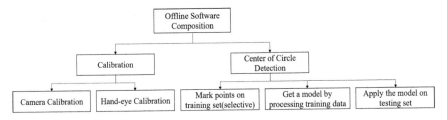

Fig. 4 Structure chart of software

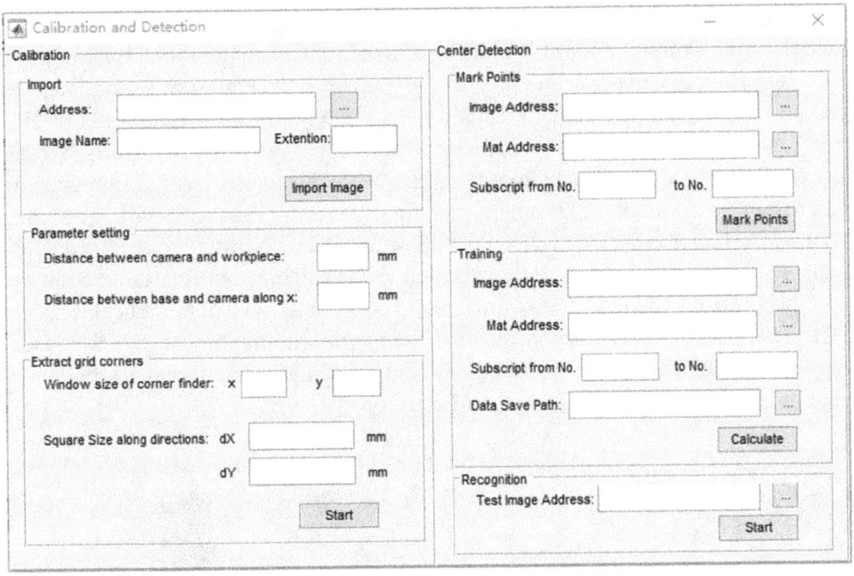

Fig. 5 Software interface

corner extractor size, and depth data which is used to transform 2D to 3D coordinate system. After obtaining the calibration board four corner pixel coordinates marked manually and depth fixed, the calibration module could work out intrinsic matrix and extrinsic matrix. Hand-eye calibration module outputs hand-eye matrix through extrinsic matrix calculated in camera calibration and pose information acquired by device itself. Calibration results are stored in.mat file.

2.2.2 Center Detection Module

Users can import new images and mark feature points to build new model. Besides, marked points were stored as.mat file. Users can also import marked.mat file directly to build model. Center coordinates data computed from model are stored in.mat file anticlockwise in order from the inside to the outside. Centers are also marked as little red circles in test image.

3 Vision System Calibration

The guidance system of this tubular heat exchanger is a triaxial flaw detection device built with stepping motor, guide rail, and strut member, as shown in Fig. 6. The triaxial device only has three degrees of freedom: translation degrees of freedom along x, y, z axes. It uses Point Grey BFLY-PGE-12A2M-CS CCD camera to collect images. Because distance between camera, detector, and workpiece has low requirement of accuracy and monocular camera cannot collect enough information on this device with three degrees of freedom, this distance has been fixed at the beginning. So that the calibration process can be simplified, 2D pixel coordinate system of image only need to be transformed to 2D camera coordinate system. The process of calculating this transform matrix is called camera calibration. Then according to the fixed depth distance, 2D coordinates can be transformed to 3D camera coordinates. In addition, hand-eye matrix which can transform camera coordinate system to tool center point (TCP) coordinate system is the result of hand-eye calibration. From this image, coordinate system can be transformed to world coordinate system so that subsequent robot guidance and path planning could carry on.

Fig. 6 Triaxial flaw detection device

3.1 Image-Forming Principle and Coordinate Transformation

Actual camera imaging is based on lens imaging pinhole model, this paper adopts an ideal pinhole model that does not consider the effects of distortion, as shown in Fig. 7. In the model, when light passes through the point P and the pinhole in 3D space at the same time, the intersection point generated on the plane behind the camera is the image of point P. The model satisfies lens imaging law:

$$\frac{1}{u} + \frac{1}{v} = \frac{1}{f} \quad (1)$$

Before calculating the transformation matrix and calibration parameters, introducing the coordinate systems used in the calibration process: Pixel Coordinate System (PCS), Image Plane Coordinate System (IPCS), Camera Coordinate System (CCS), World Coordinate System (WCS), and Tool Coordinate System (TCS). The direction of CCS and WCS is shown in Fig. 8.

PCS is in pixels; the origin of it is in the upper left corner of the imaging target surface. Pixel coordinate is expressed as $(u, v)^T$. IPCS is in physical distance mm; the origin of it is at the intersection (u_0, v_0) of the optical axis and the plane which is called principal point. The x and y axes are parallel with u-axis and v-axis, respectively, and the expression of coordinates is (x, y). The transformation from PCS to IPCS is:

$$u = x/\mathrm{d}x + u_0 \quad (2)$$

$$v = y/\mathrm{d}y + v_0 \quad (3)$$

The transformation from 2D coordinates in IPCS to 3D coordinates in CCS satisfies the imaging model above:

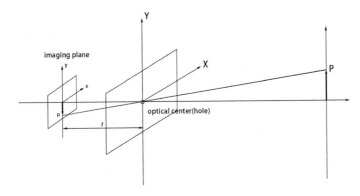

Fig. 7 Lens imaging pinhole model

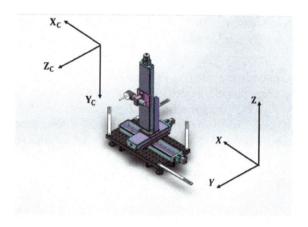

Fig. 8 Direction of CCS and WCS

$$\frac{x}{X_C} = \frac{y}{Y_C} = \frac{f}{Z_C} \tag{4}$$

In this expression, f is the distance from the focal point to the plane of the IPCS which is called focal length.

WCS is the real coordinate system in physical sense. This paper uses the flaw detection device base coordinate system as WCS. The origin of CCS is at the camera optical center and z-axis is parallel to the optical axis from the image (calibration board plane) to camera. The relative positional relationship between WCS and CCS is as shown in Fig. 8. Because workpiece plane and calibration board plane are normal to device base in this project, it is impossible to assume that the world coordinates z of points on the calibration board are 0. So conventional camera calibration methods cannot be used in this project.

3.2 Camera Calibration

The purpose of camera calibration is to calculate the camera parameters: intrinsic matrix and extrinsic matrix. The intrinsic matrix includes the physical characteristics of camera, and the extrinsic matrix is related to the shooting angle and position of camera.

Because the initial distance between CCD and calibration board is known and fixed and the detection device does not have rotation degrees of freedom, 3D world coordinate system is simplified to 2D coordinate system on the calibration board plane. Then point on calibration board $M_{2d} = [X_C, Y_C]^T$ (homogeneous form $\tilde{M}_{2d} = [X_C, Y_C, 1]^T$) is projected onto point $m = [u, v]^T$ (homogeneous form $\tilde{m} = [u, v, 1]^T$) in 2D pixel coordinate system. The transformation is:

$$s\tilde{m} = AH_1\check{M}_{2d} \tag{5}$$

In this expression, A is intrinsic matrix, H_1 is extrinsic matrix.

Apparently, there is a homography relationship between image plane and calibration board plane in 2D CCS. Because $\tilde{m}$ and $\check{M}_{2d}$ are both in homogeneous form and the form of H_1 obtained by converting between 2D coordinate system is special, it is easy to infer that $s = 1$. Among the parameters to be solved, extrinsic expresses the position of camera. Extrinsic matrix transforms corresponding CCS to initial CCS whose optical axis passes through IPCS origin. That is to say, the $x - y$ plane of WCS whose origin is base center coincides with $x_c - z_c$ plane of initial CCS. Intrinsic matrix is the embodiment of physical characteristics of camera that transform IPCS to initial CCS. Intrinsic matrix and extrinsic matrix are both calculated from 2D information:

$$A = \begin{bmatrix} \alpha & & u_0 \\ & \beta & v_0 \\ & & 1 \end{bmatrix} \tag{6}$$

$$H_1 = \begin{bmatrix} R_1 & t_1 \\ 0 & 1 \end{bmatrix} \tag{7}$$

In these expressions, α, β: scale factor related to focal length along u-axis and v-axis, respectively; u_0, v_0: coordinate of principal point on image plane; R_1: 2×2 rotation matrix (unit matrix); t_1: translation vector $\left(t_1 = \begin{bmatrix} a \\ b \end{bmatrix}\right)$.

During camera calibration, CCD shifts on $x-y$ plane and takes photographs at different positions. When treating WCS as reference system, CCS changes continuously with the shooting position. In this paper, the triaxial device does not have rotation degrees of freedom; rotation matrix R_1 is unit matrix; and the position of CCD in $x-z$ plane of WCS decides the translation vector t_1.

A H_1 transforms point m in 2D PCS to point $\check{M}_{2d}$ in 2D CCS. It is known that fixed distance between camera and calibration board is z_c. $\check{M}_{2d}$ left multiplies matrix Z can get point $M_C = (X_C, Y_C, Z_C)^T$ in 3D CCS, then subsequent hand-eye calibration can carry on:

$$Z = \begin{bmatrix} 1 & 0 & 0 \\ 0 & 1 & 0 \\ 0 & 0 & Z_c \end{bmatrix} \tag{8}$$

Transformation from point M_C (homogeneous form $\tilde{M}_c = (X_C, Y_C, Z_C, 1)^T$) in 3D CCS to point M (homogeneous form $M = (X, Y, Z, 1)^T$) in WCS is:

$$M_C = H_2 M \tag{9}$$

$$H_2 = \begin{bmatrix} R_2 & t_2 \\ 0 & 1 \end{bmatrix} \tag{10}$$

In this transformation, three axes of CCS are parallel to those of WCS, respectively, after rotating CCS 90° counterclockwise around X_C axis. Translation vector t_2 is calculated by H_1 and fixed distance dis_bc between CCD and device base along Y-axis:

$$R_2 = \begin{bmatrix} 1 & & \\ & 1 & \\ & & -1 \end{bmatrix} \tag{11}$$

$$t_2 = \begin{bmatrix} -a \\ \text{dis\_bc} \\ b \end{bmatrix} \tag{12}$$

Then, extrinsic matrix is $H = H_2 Z H_1$. Extrinsic matrix H_i of image i is the transform matrix of corresponding CCS_i and WCS.

3.3 Hand-Eye Calibration

The hand-eye calibration matrix completes the transformation from camera coordinate system to tool coordinate system. In the flaw detection device, the detector is always relatively stationary with CCD so that a typical "eye in hand" system can be formed. The 4 × 4 transformation matrix between CCS and TCS is a constant.

Hand-eye calibration needs at least three equations. When moving TCP and CCD more than twice, more than three photographs will be taken at different positions. With photographs and TCP poses information read from controller, hand-eye matrix can be calculated as the following expression (under normal conditions multiple sets of photographs will be taken):

$$T_{\text{tcp}} X = X T_{\text{ccd}} \tag{13}$$

In this expression, T_{tcp}: transformation matrix between TCS moved before and after; X: constant transformation matrix between CCS and TCS for; T_{ccd}: transformation matrix between CCS moved before and after.

4 Image Processing Algorithm

Considering the special shape of TTS weld, Hough transform was first utilized to detect circles to acquire three parameters of circles: center coordinates (x, y) and radius [7]. Then drawing the circles detected and circle centers in the image. However, because the CCD lens is too small compared to the entire workpiece plane, all other tubes in photographs appear elliptical to varying degrees besides the tube that the lens is facing. For Hough transform, the calculation of five parameters is too large to detect ellipses. In addition, thickness of workpiece greatly affects the accuracy of Hough algorithm for circular detection. In conclusion, in this paper, inverse compositional active appearance model algorithm which can extract image features and identify deformable objects excellently is adopt to extract the center of TTS weld [8].

4.1 Manually Mark Feature Points on Training Images

Marking feature points manually on the workpiece by visual inspection. Red feature points mean that position of the point on workpiece cannot change casually among training images. In a training image, marking some feature points along the image feature optionally between two red feature points can help identify deformed image feature. Distinguishing these points in blue which will redistribute according to image processing by algorithm in subsequent calculation. When marking finished, a path will form in the order of marking as shown in Fig. 9. In all training images, the order of red feature points must be the same. Apparently, the position and number of red feature points cannot be changed. In this paper, every training image includes 11 red feature points.

4.2 AAM Fitting Algorithm Based on Inverse Compositional Algorithm

4.2.1 Active Appearance Models (AAM)

Making Statistics of Shape Information
We extract all the coordinates of feature points v which make up shape vector s:

$$s = (x_1, y_1, x_2, y_2, \ldots, x_v, y_v)^T \tag{14}$$

Shape s is the base shape s_0 plus the combination of n shape vectors s_i with linear shape variation that AAMs allow as the following expression shows. In this

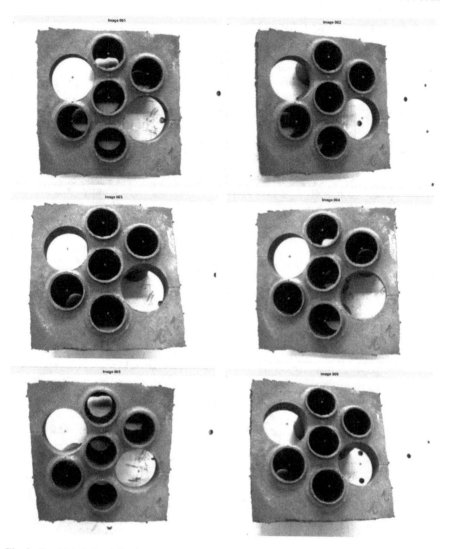

Fig. 9 Hand labeled training images

expression, p_i is shape coefficient, s_0 is mean shape, and s_i are eigenvectors corresponding to the largest eigenvalues:

$$s = s_0 + \sum_{i=1}^{n} p_i s_i \qquad (15)$$

Feature points make up meshes. Original AAMs usually use Procrustes analysis on training meshes to remove variation caused by global shape normalizing transformation. Then Principle Component Analysis (PCA) is adopt to get a result only related to local non-rigid shape deformation.

Making Statistics of Appearance Information

Similarly, appearance $A(x)$ is the base appearance $A_0(x)$ plus the combination of m appearance images $A_i(x)$. In this expression, λ_i is appearance coefficient, A_0 is mean appearance, and A_i are eigenvectors corresponding to the largest eigenvalues:

$$A(x) = A_0(x) + \sum_{i=1}^{m} \lambda_i A_i(x), \quad \forall x \in s_0 \tag{16}$$

AAMs wrap the training meshes first in accord with base mesh s_0. Then A_0 and A_i are formed from PCA analysis of training meshes. Actually, for AAMs, appearance image $A(x)$ is an image defined in base shape s_0.

Build Model Instance

Shape vectors s are formed by shape coefficient p and appearance $A(x)$ defined in base shape s_0 is formed by appearance coefficient λ in AAMs. Warping appearance A from base mesh s_0 to shape s to accomplish a piecewise affine transformation, then an AAM model instance with shape coefficient p and appearance coefficient λ is built. The result of piecewise affine warp on pixel x in s_0 is $W(x; p)$, namely the pixel in s according to x. Executing piecewise affine warp on A from s_0 to s, then the 2D image of model instance is created. The process can be expressed as the following equation:

$$M(W(x; p)) = A(x) \tag{17}$$

Fitting of Model Instance and Test Image

Calculating the difference between model instance created by AAM and test image $I(x)$. The fitting process minimize $\sum_{x \in s_0} E(x)^2$ and squares of the difference. The error between $I(x)$ and $M(W(x; p))$ at pixel x is:

$$E(x) = A_0(x) + \sum_{i=1}^{m} \lambda_i A_i(x) - I(W(x; p)) \tag{18}$$

Executing piecewise affine warp W on every pixel x in base mesh s_0 to calculate the corresponding pixel $W(x; p)$. Pixels $W(x; p)$ make up sample $I(W(x; p))$ corresponding to input image. The difference between appearance $A_0(x) + \sum_{i=1}^{m} \lambda_i A_i(x)$ and $I(W(x; p))$ is error image E.

The Application of Inverse Compositional Algorithm in AAMs

After executing affine transformation $W(x; p)$, global shape normalization $N(x; p)$ will be adopt on image. In this expression, (a, b) is calculated by ratio k and rotation angle θ and (t_x, t_y) is transformed from coordinate (x, y):

$$N(x; p) = \begin{pmatrix} (1+a) & -b \\ b & (1+a) \end{pmatrix} \begin{pmatrix} x \\ y \end{pmatrix} + \begin{pmatrix} t_x \\ t_y \end{pmatrix} \qquad (19)$$

The final fitting result M is calculated through $W(x; p)$ and $N(x; p)$ by AAM fitting algorithm based on inverse compositional algorithm. Firstly, executing affine transformation on appearance A based on base mesh s_0 to image $W(x; p)$. Then executing normalization $N(x; p)$ on it. Finally through warping, 2D image M with appropriate size and shape that includes model instance is created:

$$M(N(W(x; p); q)) = A(x) = A_0(x) + \sum_{i=1}^{m} \lambda_i A_i(x) \qquad (20)$$

4.3 Image Detection Experiments

After inputting hand labeled feature points matrix and original image together into algorithm, algorithm divides red feature points in every image into three parts: center of the central tube, centers of surrounding six tubes, and four corners of the workpiece. The three parts formed three closed polygons with blue feature points between red points, respectively. Apparently, there is no redistributed blue feature points between centers of central tube and the above tube. The center of the above tube and the upper left corner are not connected in a similar way. The three parts are warped independently so that the result is a better fit to the image. The general center position is asked to choose before wrapping as shown in Fig. 10.

As shown in Fig. 11, during the fitting of model instance and test image, the coincidence degree of feature points in model instance and that in image continues to grow with iteration from (a) to (d). Final center detection result is shown in Fig. 12.

Fig. 10 Contour position selection

A Method for Detecting Central Coordinates of Girth Welds ... 79

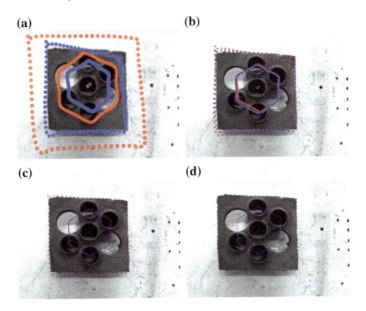

Fig. 11 Wrapping process

Fig. 12 Calculated result of grid weld centers

5 Conclusion

In view of the complexity and difficulty of TTS welding process, flaw detection is an important flow during tubular heat exchanger production process. To improve the degree of automation and reduce the time and labor cost, this paper studies a method which can realize automatic center recognition. In particular, the following conclusions can be made:

1. Based on CCD camera and self-designed triaxial motion device without rotational degrees of freedom, hardware and software systems were developed to acquire center coordinates of TTS weld. The system includes calibration module and center detection module.
2. Focusing on calibration based on motion device without rotational degrees of freedom and extraction of TTS center coordinates, a method which integrates AAM fitting algorithm and inverse compositional algorithm is proposed. Besides, in consideration of the self-designed motion device, the conventional calibration algorithm was simplified to fit it.
3. Because the triaxial flaw detection device only has three translational degrees of freedom, distance along optical axis between CCD camera, workpiece, and base center can be fixed so that camera calibration can be simplified.

Under the circumstance of low requirement of accuracy of center coordinates, inverse compositional AAM can identify center of grid weld through appropriate labeled training set. Then, the ellipse identification problem caused by shooting angle and position of monocular camera can be solved. The center detection result is accepted within the margin of error by visual inspection.

References

1. Xu S, Wang W (2013) Numerical investigation on weld residual stresses in tube to tube sheet joint of a heat exchanger. Int J Press Vessels Pip 101(7):37–44
2. Cui R (2000) Quality control of tube-to-sheet joint on tubular heat exchanger. Shanxi Chemical Industry
3. Jin Z, Li H, Zhang C et al (2017) Online welding path detection in automatic tube-to-tubesheet welding using passive vision. Int J Adv Manuf Technol 90(9–12):3075–3084
4. Qiang TP, Wei XU, Chen YK et al (2004) One special technique of non-destructive testing-radiographic testing on tube to tube-sheet welds. Press Vessel Technol
5. Arunmuthu K, Saravanan T, Philip J et al (2008) Image processing of radiographs of tube-to-tubesheet weld joints for enhanced detectability of defects. Insight Non-Destr Test Cond Monit 50(6):298–303
6. Ke-Qin D, Guang C, Xu Z (2012) Design of ray digital scanning device for the tube-to-tube sheet fillet. Mach Des Manuf
7. Basca CA et al (2006) Randomized hough transform for ellipse detection with result clustering. In: The international conference on "computer as a tool", vol 2. IEEE, Belgrade, pp 1397–1400
8. Matthews I, Baker S (2004) Active appearance models revisited. Int J Comput Vision 60(2):135–164

Yu Ge was born in China in 1995. She received a bachelor degree in materials processing engineering from Wuhan University of Technology, Wuhan, China, in 2017. She studies in intelligentized robotic welding technology laboratory for a master's degree in Shanghai Jiao Tong University since September 2017 until now. Her research directions are machine vision and wire and arc additive manufacturing.

Yanling Xu was born in China in 1980. He received the Ph.D. degree in materials processing engineering from Shanghai Jiao Tong University, Shanghai, China, in 2012. Since April 2014, as a lecturer, he works in Shanghai Jiao Tong University, Shanghai, China, intelligentized robotic welding technology laboratory. He mainly engaged in researches in robotic welding automation technology, machine vision sensing technology, and wire and arc additive manufacturing.

Spectral Signal Analysis Using VMD in Pulsed GTAW Process of 5A06 Al Alloy

Haiping Chen, Gang Li, Na Lv and Shanben Chen

Abstract Spectral signal collected during GTAW process is multi-dimensional, thus difficult to be analyzed. In previous researches, SOI and EMD algorithms were used to reduce dimensionality. In this paper, an automatic discriminant criterion based on correlation coefficient is proposed to eliminate redundant wavelength signals in spectral domain (200–1100 nm), and only a few spectral lines will be left for subsequent processing. To overcome the limit of EMD, variational mode decomposition (VMD) algorithm is used to decompose the spectral signal into determined number of intrinsic signals with fewer modal aliasing in time domain. The number of VMD modes is nine determined by the spectral peaks' number in frequency domain. Finally, it has been proved that VMD decomposed models have more physical meaning and are more feasible for real-time analysis in compared with EMD algorithm.

Keywords Arc spectrum · Feature selection · Variational mode decomposition · GTAW

1 Introduction

Porosity is a kind of defects forming inside seam during welding process. When the porosities occur, the quality of the weldment will be influenced and may no longer meet relevant mechanical indexes [1]. To ensure the performance of the weld, non-destructive testing is usually used for the weld seam after welding. At present, non-destructive testing methods commonly used include X-ray detection [2], ultrasonic detection [3], and spectral detection [4], etc.

Dinda et al. [5] used X-ray tomography to 3D image and quantify porosity in electron bean welding. Tao et al. [6] employed an acoustic emission technical to detect porosity based on the characteristic parameters of signals such as amplitude

---

H. Chen · G. Li · N. Lv · S. Chen (✉)
School of Materials Science and Engineering,
Shanghai Jiao Tong University, Shanghai 200240, China
e-mail: sbchen@sjtu.edu.cn

and centroid frequency. However, above methods have higher requirements on equipment and instruments. An effective way with low cost is expected to develop. The arc spectrum contains information on metal vapors, shielding gases, and arc gases. Therefore, it has an intrinsic connection with the internal defects of the weld, which makes it possible to become a new method of real-time detection.

Sibillano et al. [7] found the relationship between plasma electron temperature and penetration using three different types of laser source and apply it to a real-time detection method for penetration. Yu et al. [8] proposed a method using the intensity ratio of H-line and Ar-line to predict the most probable positions of porosity. Huang et al. [9] used the empirical mode decomposition (EMD) algorithm to decompose the ratio signals and determine the low-frequency mode as the key to detect the presence of porosity. And to reduce the dimensionality in spectrum, principal component analysis (PCA) was used to extract the principal components of spectrum of interests (SOI) [10].

In this paper, an automatic discriminant criterion based on Pearson's correlation coefficient is proposed to eliminate redundant wavelength instead of selecting the SOI manually. A few spectral lines can be selected to represent the whole spectrum. Then, an analysis based on VMD is performed on the intensity series in time domain. Compared with the EMD, it is concluded that VMD modes have more physical meaning and more suitable for real-time analysis.

2 Experiment

2.1 Experimental System

The experimental system is mainly composed of a welding system and a spectrum acquisition system. For welding system, it mainly includes a robot system, a welding machine, a welding gun, a wire feeder, a shielding gas device, and a water-cooling device. And for spectrum acquisition system, it mainly includes ND filter, optical fiber, spectrometer, and related software. The organizational chart of experimental system is shown as Fig. 1.

2.2 Experimental Design

In this experiment, 4 mm-thick 5A06 aluminum alloy plate was used as the base metal, while ER5183 as welding wire. Their component elements are shown in Table 1.

The 99.9% argon was used as shielding gas with a flow rate of 14 L/min. The tungsten electrode had a diameter of 3.2 mm and adds 2% ThO_2. The other important welding parameters are shown in Table 2. The peak current takes a segmentation reduction from 240 to 210 A.

Spectral Signal Analysis Using VMD in Pulsed GTAW Process ...

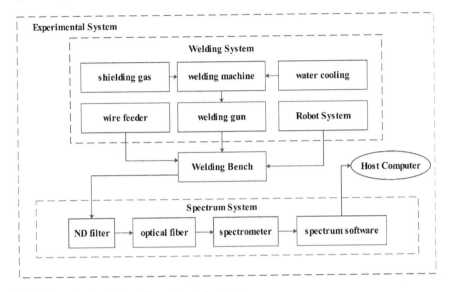

Fig. 1 Organizational chart of experimental system

Table 1 Component table of 5A06 Al alloy and ER5183 welding wire

Element	Al	Mg	Mn	Fe	Si	Zn	Cu
5A06	Bal.	5.8–6.8	0.5–0.8	≤0.4	≤0.4	≤0.2	≤0.1
ER5183	Bal.	4.3–5.2	0.5–1.0	≤0.4	≤0.4	≤0.25	≤0.1

Real-time spectrum data were acquired using a commercial spectrometer Ocean Optics HR4000. The effective spectrum range is from 200 to 1100 nm with a resolution of 0.2 nm. The sample period was set to 30 ms and the intensity of signal was set from 0 to 16,000. The whole spectrum data acquired in the experiment are the changes in the spectrum that are shown in Fig. 2, where x-axis represents different wavelengths, y-axis represents time (in this case, 1 s corresponds to about 33 sampling points because of sample period) and z-axis represents intensity of each sample points at a certain wavelength.

Table 2 Welding parameters

Pulse frequency	Pulse duty ratio	Base current	Peak current	Welding speed	Feeding speed
1 Hz	40%	40 A	240–210 A	3 mm/s	10 mm/s

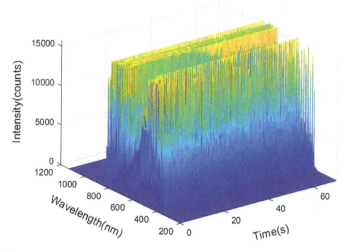

Fig. 2 Full spectrum data during GTAW of 5A06 Al alloy

3 Theory

3.1 Automatic Discriminant Criterion for Classification

The Pearson's correlation coefficient is a statistic used to reflect the degree of linear correlation between two variables which can be assumed by Eq. (1). Each wavelength with a signal series in time domain can be seen as an independent variable, for which a Person correlation coefficient matrix is calculated. As shown in Fig. 3, most of spectral line signals are strongly linearly correlated. It is obvious that there is a lot of redundant information in the spectrum data.

$$r = \frac{\sum_{i=1}^{n}(X_i - \overline{X})(Y_i - \overline{Y})}{\sqrt{\sum_{i=1}^{n}(X_i - \overline{X})^2}\sqrt{\sum_{i=1}^{n}(Y_i - \overline{Y})^2}} \quad (1)$$

As to reduce the size of data in spectrum domain, a proper spectral line can be selected to represent a group of strong linear correlation dataset. It is considered that classification for spectrum should satisfy two rules. The first rule is the correlation coefficient of variables in a same type should be greater than the threshold. The second rule is the correlation coefficient of any two variables in different types, respectively, is smaller than threshold. The algorithm procedure is described as following:

1. Initialize each wavelength as different types and set the threshold value.
2. Choose the maximum value of correlation coefficient matrix between two different types and combine them.
3. Update the coefficient matrix.

Spectral Signal Analysis Using VMD in Pulsed GTAW Process ...

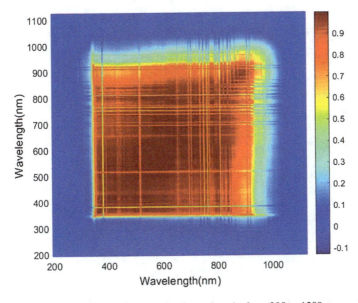

Fig. 3 Pearson's correlation coefficient matrix of wavelengths from 200 to 1200 nm

4. End the procedure if the maximum value of matrix is smaller than threshold, otherwise return to step 2.

3.2 Variational Mode Decomposition

Variational mode decomposition (VMD) is non-recursive algorithm proposed by Dragomiretskiy and Zosso [11], which is used to decompose nonlinear, non-stationary signals into intrinsic mode signals with limited bandwidth. VMD algorithm solves the problem of uncertain number of modes and modal aliasing which caused by EMD. The algorithm procedure is described as following:

- Initialize mode function $\{u_k^1\}$, center frequency $\{\omega_k^1\}$, and Lagrange multiplier λ^1.
- Iterative update the value of u_k^i, ω_k^i, and λ^i using Eqs. (2)–(4).
- Stop after satisfy the criteria, and u_k^n is the k decomposed model.

$$u_k^{n+1}(\omega) = \frac{f(\omega) - \sum_{i<k} u_k^{n+1}(\omega) - \sum_{i>k} u_k^{n+1}(\omega) + \frac{\lambda(\omega)}{2}}{1 + 2\alpha(\omega - \omega_k^n)^2} \quad (2)$$

$$\omega_k^{n+1} = \frac{\int_0^\infty \omega |u_k^{n+1}(\omega)|^2 d\omega}{\int_0^\infty |u_k^{n+1}(\omega)|^2 d\omega} \quad (3)$$

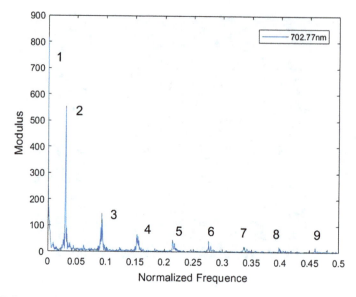

Fig. 4 Unilateral spectrum of 702.77 nm signals

$$\lambda^{n+1}(\omega) = \lambda^{n}(\omega) + \tau \left(f(\omega) - \sum_{k} u_{k}^{n+1}(\omega) \right) \qquad (4)$$

The number of decomposed models for VMD k is a key parameter needed to be set in advance. Fourier transform is performed on a signal at a certain wavelength (e.g. 702.77 nm), and the corresponding unilateral spectrum is shown in Fig. 4. It is clear that signals mainly focus on nine peaks. So the number of peak in unilateral spectrum is used as the number of decomposed models by VMD. In this paper, the k is set to 9.

4 Results and Discussion

4.1 The Classification of Spectrum

According to the correlation coefficient matrix, it is found the spectrum from 200 to 350 nm and 900 to 1100 are invalid data. To classify 350–900 nm spectrum, it is needed to set threshold value at first. The different classification results based on threshold is shown in Fig. 5. The index of types is named by descend order of number of wavelengths in the types, which means Type 1 has the maximum number of wavelengths. It is shown that Type 1 and Type 2 include more than 60% of the spectrum. And as the threshold increasing, proportion of quantity of Type 1 is decreasing while

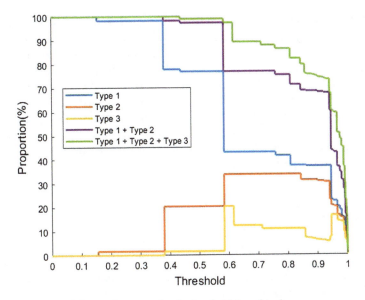

Fig. 5 Classification of the top three types by the threshold from 0 to 1

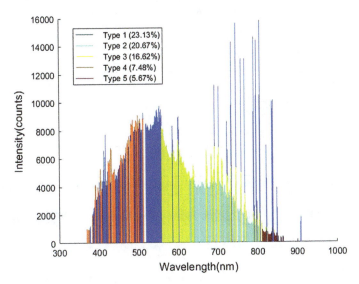

Fig. 6 Classification of top five types and corresponding proportions

proportion of quantity of Type 2 is increasing. It is interpreted as strict classification criterion (higher threshold value) makes them hard to be considered as the same type. 0.946 is set to threshold value because the proportion of top three types is greater than 60% with more evenly distribution.

Partial classification is shown in Fig. 6 at threshold 0.946. It shows that the signals of nearby wavelengths tend to be the same type because they have highly similar variations. And the spectral characteristic lines from the same element are easily classified as the same class, looking at the characteristic lines from 700 to 850 nm which are caused by Ar. The top five types consist of 73.57% spectrum shown in Fig. 6. Moreover, the top 10 types and 15 types can be related to 87.47 and 90.46% spectrum, respectively. It supports the reduction method using a few spectral signals to represent full spectrum dataset. It is an effective and useful way to significantly reduce redundancy in spectrum domain. For the future analysis, some spectral lines will be selected as spectrum feature applied to various models.

4.2 The VMD Modes

In time domain, a spectral signal is interfered by many other signals such as background signal and pulse signal. In order to get a clearer signal, VMD algorithm is applied to the spectral signal. As mentioned above, the number of modes k is set to 9 in this paper. The decomposition results at 702.77 nm are shown in Fig. 7. It is found that the overall change trend of Mode 1 is similar to the original signal and the

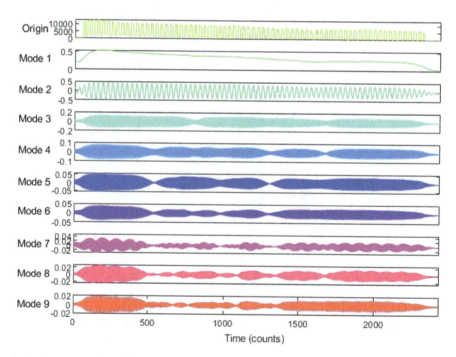

Fig. 7 Decomposed models for VMD at 702.77 nm

Spectral Signal Analysis Using VMD in Pulsed GTAW Process ...

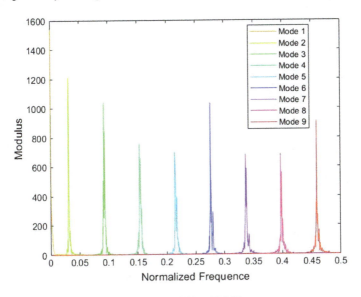

Fig. 8 Unilateral spectrum for nine modes by VMD at 702.77 nm

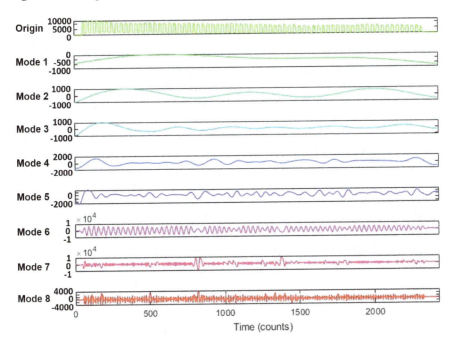

Fig. 9 Decomposed models for EMD at 702.77 nm

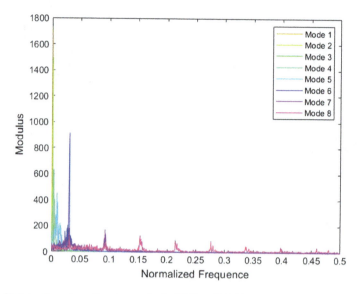

Fig. 10 Unilateral spectrum for nine modes by EMD at 702.77 nm

Mode 2 is similar to pulse signal. Although it is hard to specific physical meaning for Mode 4–9, it is still obvious that they have same characteristics and may be caused by same factors. Each mode is transferred to unilateral spectrum and shown in Fig. 8. Compared with unilateral spectrum of the original signal, it is clear that VMD can result in good decomposed models with narrow bandwidth because peaks of them can correspond one-to-one.

It is concluded that the VMD can well separate the interference signals in the spectral signal, and some of modes have a clear physical meaning.

The decomposed modes and related unilateral spectrum by EMD are shown in Figs. 9 and 10. Compared with VMD, the decomposed modes by EMD have a large frequency range and modal aliasing with less physical meaning. And EMD uses an adaptive k according to the length of signal, which brings uncertainty to the real-time analysis because the existence of a mode is unknown. As a result, signal decomposition using VMD is more advantageous than EMD, both in terms of signal interpretation and subsequent real-time analysis processing.

5 Conclusion

In this paper, spectral signals are analyzed in spectrum domain and time domain, respectively. Firstly, in spectrum domain, an automatic discriminant criterion for classification based on correlation coefficient is proposed to eliminate redundant wavelength signals. A few spectral lines can represent the whole spectrum. Secondly, in time domain, VMD algorithm is applied to the spectral signal and it is considered that Mode 1 represents the overall tendency and Mode 2 represents the pulse signal. The number of modes can be determined by the number of peaks in frequency domain. And compared with EMD, VMD have narrower bandwidth, more physical meaning and are more feasible for real-time analysis.

Acknowledgements This work is supported by the National Natural Science Foundation of China (51575349).

References

1. Ascari A, Fortunato A, Orazi L et al (2012) The influence of process parameters on porosity formation in hybrid LASER-GMA welding of AA 6082 aluminum alloy. Opt Laser Technol 44(5):1485–1490
2. Yamamoto S, Hoshi T, Miura T et al (2014) Defect detection in thick weld structure using welding in-process laser ultrasonic testing system. Mater Trans 55(7):998–1002
3. Zou Y, Du D, Chang B et al (2015) Automatic weld defect detection method based on Kalman filtering for real-time radiographic inspection of spiral pipe. NDT E Int 72:1–9
4. Harooni M, Carlson B, Kovacevic R (2014) Detection of defects in laser welding of AZ31 B magnesium alloy in zero-gap lap joint configuration by a real-time spectroscopic analysis. Opt Lasers Eng 56:54–66
5. Dinda SK, Warnett JM, Williams MA et al (2016) 3D imaging and quantification of porosity in electron beam welded dissimilar steel to Fe-Al alloy joints by X-ray tomography. Mater Des 96:224–231
6. Tao Y, Wang W, Sun B (2014) Nondestructive online detection of welding defects in track crane boom using acoustic emission technique. Adv Mech Eng 6:505464
7. Sibillano T, Rizzi D, Ancona A et al (2012) Spectroscopic monitoring of penetration depth in CO_2 Nd:YAG and fiber laser welding processes. J Mater Process Technol 212(4):910–916
8. Yu H, Xu Y, Song J et al (2015) On-line monitor of hydrogen porosity based on arc spectral information in Al–Mg alloy pulsed gas tungsten arc welding. Opt Laser Technol 70:30–38
9. Huang Y, Wu D, Zhang Z et al (2017) EMD-based pulsed TIG welding process porosity defect detection and defect diagnosis using GA-SVM. J Mater Process Technol 239:92–102
10. Huang Y, Zhao D, Chen H et al (2018) Porosity detection in pulsed GTA welding of 5A06 Al alloy through spectral analysis. J Mater Process Technol 259:332–340
11. Dragomiretskiy K, Zosso D (2014) Variational mode decomposition. IEEE Trans Sig Proc 62(3):531–544

Haiping Chen received his B.S. degree in School of Materials Science and Engineering (SMSE) from Shanghai Jiao Tong University in 2016, and then he continued to study for an M.S. degree in Intelligentized Robotic Welding Technology Laboratory in SMSE, SJTU. His interests are robotic welding, feature selection and extraction, spectral analysis and real-time control during welding process.

Prof. Shanben Chen received his B.S. degree in industrial automation from Dalian Railway Institute (Dalian Jiao Tong University) in 1982, and received his M.S. and Ph.D. in control theory and application from Harbin Institute of Technology, China, in 1987 and 1991, respectively. From 2000 to present, he has served as the Special Professor, Cheung Kong Scholar Program of the Ministry of Education of China & Li Ka Shing Foundation, Hong Kong, and engaged at Shanghai Jiao Tong University, China, where he is also director of the Intelligentized Robotic Welding Technology Laboratory. Prof. Chen has also been a visiting professor at the University of Western Sydney (UWS) in connection with the ARC Linkage collaboration since 2009.

Nonlinear Identification of Weld Penetration Control System in Pulsed Gas Metal Arc Welding

Wandong Wang, Zhijiang Wang, Shengsun Hu, Yue Cao and Shuangyang Zou

Abstract Weld penetration plays an important role in the joint strength and its control has always been the focus of study. The paper established a single-input–single-output (SISO) weld penetration control system in pulsed gas metal arc welding (GMAW-P), where the base current (I_b) was taken as system input and the change in arc voltage during peak current period (ΔU) was taken as the system output. According to the nonlinear relationship between I_b and ΔU, a Hammerstein model with disturbances, composed of nonlinear static model and linear dynamic model, was proposed to describe the nonlinear control system. The nonlinear static system was determined based on the model of I_b and ΔU in steady state, and the parameters of linear dynamic system were identified by the recursive least square algorithm. Pseudo-random ternary signals (PRTS) were designed for the system identification. The identified results showed the Hammerstein model with disturbances can represent the penetration control system in GMAW-P within an acceptable range, which was validated by the step experiments data.

Keywords Hammerstein model · System identification · Weld penetration control · Pulsed gas metal arc welding

1 Introduction

Weld penetration is one of the most important factors to determine the mechanical strength of weld in pulsed gas metal arc welding (GMAW-P), which is widely used in manufacturing field for its versatile and easily automated advantages [1, 2]. With the development of intelligent manufacturing technologies, the real-time control of weld

W. Wang · Z. Wang (✉) · S. Hu · Y. Cao · S. Zou
Tianjin Key Laboratory of Advanced Joining Technology, School of Materials Science and Engineering, Tianjin University, Tianjin 300350, China
e-mail: wangzj@tju.edu.cn

W. Wang
e-mail: wangwandong@tju.edu.cn

penetration has become a research focus. However, the weld penetration cannot be measured directly during welding process. Hence, many new sensing methods were developed to measure the dynamic characteristics of weld pool surface to determine the real-time changes in penetration [3]. Among the methods proposed in the existing literature, pool oscillation sensing [4], infrared sensing [5] and visual sensing [6] methods received most attention and were widely studied. During a welding process, arc voltage is closely to the oscillation frequency [7] and oscillation amplitude of weld pool surface, thus, the arc voltage sensing method was widely studied and applied in the control of weld penetration for its easy accessibility and better real-time performance.

Wang et al. [1, 8, 9] proposed that the change in arc voltage during peak current period (ΔU) can provide an accurate prediction for the depth of weld penetration during GMAW-P which was closely related to the oscillation amplitude of weld pool. The studies have shown that the control of weld penetration can be achieved by controlling the amplitude of voltage variation during the welding process, which can be realized by regulating pulse current parameters in real time. Hence, it is critical to establish a suitable model between pulse current parameters and characteristic variables, which can reflect weld penetration in real time, so that the penetration control system can be designed more stably and accurately. Currently, Liu et al. established a nonlinear dynamic model to correlate the process inputs and weld penetration in gas tungsten arc welding (GTAW) process which was better than dynamic linear model and presented a predictive control system of penetration based on nonlinear neuro-fuzzy model [10–12]. GMAW-P process is more complicated than GTAW process for droplet transfer, thus, a nonlinear system model should be considered to realize more accurate control when establishing the mathematical model between the input of pulsed current parameters and the output of weld penetration. Among pulse parameters, the peak parameters should be kept as constant for the peak current and peak current period will influence the accuracy of ΔU, thus, the adjustable parameters left are base current (I_b) and base current time (t_b). The change in t_b will cause the variation of pulse frequency, which is closely related to weld surface shaping, especially the uniformity of ripples. Hence, I_b, which has less influence on weld pool surface shaping, was selected as system input. This work proposes to adjust base current of pulse parameters (I_b) in real time to make ΔU fluctuate near the value of stability during the GMAW-P process, so that the predicted weld penetration can be controlled indirectly. A Hammerstein model is proposed to describe the relationship between I_b and ΔU for the weld penetration control in GMAW-P, which is a single-input–single-output (SISO) system with the base current as system input and ΔU as the system output.

As mentioned above, the identification of Hammerstein model established is the key to realize the control of weld penetration. A majority of nonlinear identification methods have already been devoted to identify the Hammerstein model [13–17]. A recursive least square identification algorithm was employed to identify the model parameters of weld penetration control system. All the experiments need to be dynamic and be conducted using step response signals and pseudo-random ternary signals (PRTS) as system inputs, which can also be used for the validation of identified model [18, 19].

2 Experimental Procedures

The GMAW-P system was established with Miller XMT 350 welding power, electrical signal acquisition system and HGT-3C workbench, as shown in Fig. 1. The sensor system was based on the principle of Hall transducers, and the sampling period was set to 0.6 ms. The travel speed of welding torch was set to 3.2 mm/s. In the experiment, the base metal used was Q235 steel plates of 270 mm × 50 mm × 4 mm, and the filler wire was MG50-6 with a diameter of 1 mm. Butt welding was conducted in the study, where "I" groove was designed and the gap between two workpieces was set to 1 mm.

In the study, a single variable, I_b, was used to achieve different weld penetrations (d_p), and the other experiment parameters were kept in constant, as shown in Table 1. I_b was set as 50, 60, 70, 80, 90, and 100 A. The relationship between ΔU and d_p was obtained and examined. And the relationship between I_b and ΔU was obtained for the identification of nonlinear weld penetration control system.

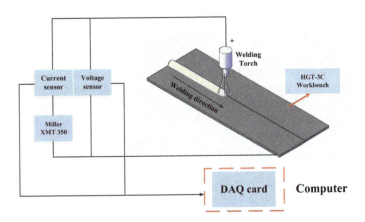

Fig. 1 Experimental setup

Table 1 Constant welding parameters of GMAW-P

Peak current (A)	Peak current period (ms)	Wire feed speed (m/min)	Base current period (ms)	Travel speed (mm/s)	Contact tip to workpiece distance (mm)	Gas flow rate (L/min)
280	54	4.8	108	3.2	15	17

3 Results and Discussion

3.1 The Characterization of Weld Penetration

The relationship between ΔU and d_p was obtained, as shown in Fig. 2. With the increase of d_p, ΔU shows a downward trend. The relationship expressed in Eq. (1) was obtained by the least square fitting and the performance of linear fitting is good from Fig. 2, which indicates ΔU has a good linear relationship with d_p. It is feasible to use ΔU as the output of weld penetration control system.

$$\Delta U = -1.27 \times d_p + 8.19 \tag{1}$$

3.2 Step Response and Steady-State Analysis

Four groups of step response experiments were designed to analyze the time-domain response of the system, and the main parameters are shown in Table 2, where the I_b (1) indicated system input in front of the step change and I_b (2) indicated system input after the step change. The results of step response experiments are shown in Fig. 3. Analyzing system input (I_b) and system output (ΔU) in Fig. 3, it can be seen that: ① the noises of the system output were quite large, and the system delay was not fixed, and the delay time was between 0.3 and 0.8 s; ② the output fluctuated in a certain range around the steady-state value, which was the average value of the system output before and after the step change.

The steady-state values of the system output obtained under different inputs of I_b are listed in Table 3. The static nonlinearity of system can be described with a fitted quadratic curve as shown in Fig. 4, expressed by Eq. (2). It showed that there is a nonlinear relationship in the weld penetration control system.

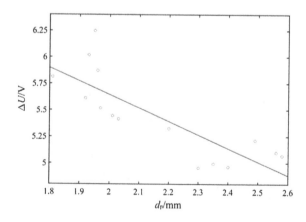

Fig. 2 The relationship between ΔU and weld penetration

Nonlinear Identification of Weld Penetration Control System ...

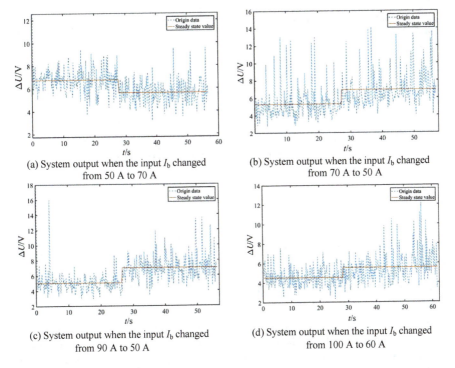

(a) System output when the input I_b changed from 50 A to 70 A

(b) System output when the input I_b changed from 70 A to 50 A

(c) System output when the input I_b changed from 90 A to 50 A

(d) System output when the input I_b changed from 100 A to 60 A

Fig. 3 Step response experiments

Fig. 4 Static nonlinearity of weld penetration control system

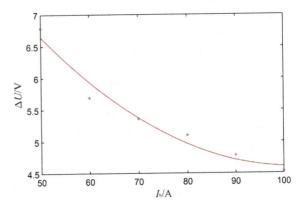

Table 2 Parameters for step experiments

Experimental No.	$I_b(1)$/A	$I_b(2)$/A
1	50	70
2	70	50
3	90	50
4	100	60

Table 3 Steady-state values for system output

I_b/A	50	60	70	80	90	100
ΔU/V	6.785	5.692	5.356	5.101	5.038	4.840

$$\Delta U = 0.00076 \times I_b^2 - 0.155 \times I_b + 12.5 \qquad (2)$$

where the ΔU is system output, and the I_b is system input.

3.3 Nonlinear Hammerstein Model

The paper proposed a Hammerstein model for identification of the nonlinear weld penetration control system. The block diagram for Hammerstein model with disturbances was shown in Fig. 5, where $u(k)$, $v(k)$, $y(k)$, and $x(k)$ are the system input, system noise, system output, and intermediate process variable, respectively. The Hammerstein model consisted of a nonlinear static model and a linear dynamic model, where $x(k)$ represents the nonlinear polynomial function and $G(z-1)$ represents the linear transfer function. $x(k)$ was expressed in Eq. (3), where the parameters are based on Eq. (2). The parameters in Eq. (2) described the relationship between $u(k)$ and $y(k)$ in steady state. In steady state of the system, there was a gain between $x(k)$ and $y(k)$. So the study described the nonlinear static system with a polynomial of variable parameters, which varied with the gain of linear dynamic system.

After several attempts to identification of linear dynamic model, third-order linear system was found to be most suitable for it, as shown in Eq. (4). The steady-state gain of the linear dynamic system can be obtained from the transfer function $G(z-1)$, as shown in Eq. (5). From Fig. 5, the residual items ($v(k)$) were introduced into the control system owing to the time-varying characteristics of the system, which can reflect the effect of noise and systematic error. Hence, according to Eqs. (3)–(5), the linear dynamic model can be expressed as Eqs. (6)–(8).

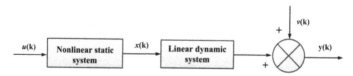

Fig. 5 Hammerstein model with disturbances [20]

$$x(k) = C_2 + C_1 \times u(k) + C_0 \times u^2(k) \tag{3}$$

where $C_0 = 0.00076/g_s$, $C_1 = -0.155/g_s$, $C_2 = 12.5/g_s$, and g_s represent the steady-state gain of linear dynamic system.

$$G(z^{-1}) = \frac{B(z^{-1})}{A(z^{-1})} = \frac{(b_0 + b_1 z^{-1} + b_2 z^{-2} + b_3 z^{-3} + b_4 z^{-4}) z^{-d}}{(1 + a_1 z^{-1} + a_2 z^{-2} + a_3 z^{-3})} \tag{4}$$

where d represents system delay, $d = 3$ in the present system.

$$g_s = \frac{b_0 + b_1 + b_2 + b_3 + b_4}{1 + a_1 + a_2 + a_3} \tag{5}$$

where g_s represents the system gain.

$$\begin{cases} y(k) = a_1 y(k-1) + a_2 y(k-2) + a_3 y(k-3) \\ \quad + b_0 x(k-d) + b_1 x(k-d-1) + b_2 x(k-d-2) \\ \quad + b_3 x(k-d-3) + b_4 x(k-d-4) + c_1 v(k-1) \end{cases} \tag{6}$$

where

$$x(k) = \frac{1}{g_s}(0.00076 \times u^2(k) - 0.155 \times u(k) + 12.5) \tag{7}$$

$$v(k) = y(k) - \hat{y}(k) \tag{8}$$

The recursive least square algorithm was used to identify the linear dynamic model, which can be described in Eq. (9).

$$\begin{cases} \hat{\theta}(k) = \hat{\theta}(k-1) + K(k)\left(y(k) - \varphi^T(k)\hat{\theta}(k-1)\right) \\ K(k) = P(k-1)\varphi(k)(\lambda + \varphi^T(k)P(k-1)\varphi(k)) \\ P(k) = \left(I - K(k)\varphi^T(k)\right)P(k-1)/\lambda \end{cases} \tag{9}$$

where

$$\varphi(k) = \begin{cases} [-y(k-1); -y(k-2); -y(k-3); x(k-d-1); \\ x(k-d-2); x(k-d-3); x(k-d-4); v(k-1)] \end{cases} \tag{10}$$

$$\theta = [a_1; a_2; a_3; b_0; b_1; b_2; b_3; b_4; c_1] \tag{11}$$

and λ is the forgetting factor, which is counter measure for the time-varying system, $\lambda = 0.95$ in the experiments; $\varphi(k)$ and θ can be described as Eqs. (10) and (11); $P(k)$, $K(k)$ and I are the standard definition in the recursive least square algorithm.

3.4 PRTS Inputs for Identification of Linear Dynamic System

PRTS is widely used in the identification of linear systems, thus, a PRTS signal with the length of 80 was selected, as shown in Fig. 6a, which was generated by solving differential equations shown in Eq. (12). The actual system input signals varied from 50 to 100 A to control weld penetration and to guarantee the stability of arc. The original PRTS signal was connected with actual welding inputs, as shown in Fig. 6b. The welding experiments were conducted with different PRTS signals as system input. Figure 7 is one of the experimental results when the system input signals were set as 50, 70 and 90 A, which will be used to identify and verify the dynamic model of weld penetration control system.

$$T_s(k) = m_1 \odot_r T_s(k-1) \oplus_r m_2 \odot_r T_s(k-2) \oplus_r m_3 \ldots \oplus_r m_n \odot_r T_s(k-n) \quad (12)$$

where $\odot_r$ is the multiplicative operator with a modulus of r, and $\oplus_r$ is the addition operator with a modulus of r; and m_i ($i = 1, 2, 3 \ldots n$) is the feedback coefficient;

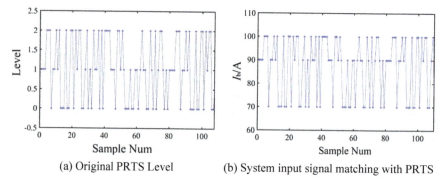

(a) Original PRTS Level (b) System input signal matching with PRTS

Fig. 6 System test signals

Fig. 7 Experimental result with PRTS as the system input

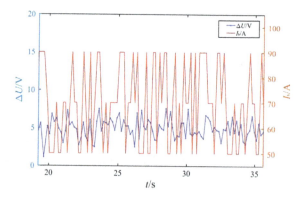

and in the present work $r = 3, n = 4, m_1 = 0, m_2 = 0, m_3 = 1, m_4 = 1, T_s(0) = 1, T_s(-1) = 0, T_s(-2) = 2, T_s(-3) = 0.$

3.5 System Identification and Model Validation

A system identification algorithm based on Eq. (9) was used in the work, whose principle can be described as Fig. 8. The results of system outputs using different PRTS as inputs were used as identification data. The model parameters are initialized to obtain the inputs of linear dynamic system ($x(k)$). The parameters of linear system are then identified by the recursive least square algorithm and the parameter of nonlinear static system is acquired by calculating the gain of linear dynamic system. When the system inputs and outputs are updated, the model parameters will also be updated. Thus, the identified model parameters will fluctuate within an acceptable range when the inputs and outputs of system reach a certain length, which means identification length.

Table 4 showed the identification results of dynamic linear system model parameters, and apparently, the system parameters were not constant. Hence, adaptive control algorithm should be added to the control system to identify the system model parameters online. The quality of model identification can be judged by relative mean square error (RMSE), as expressed in Eq. (13). Figure 9 showed the output of

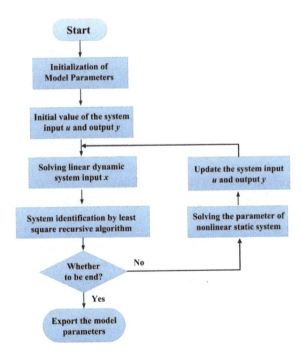

Fig. 8 Flow chart of system identification algorithm

Table 4 Identified results of dynamic linear system in different PRTS experiments

No.	a_1	a_2	a_3	b_0	b_1	b_2	b_3	b_4	c_1
1	0.2686	−0.2795	−0.1618	−0.6357	−1.004	1.1231	1.1248	0.2685	0.3102
2	−0.0849	−0.0238	0.1758	0.0549	0.0366	−0.0206	0.1152	−0.0593	0.2556
3	−0.1831	0.0020	−0.2723	0.1230	0.2331	−0.1566	0.0258	0.1028	−0.2203

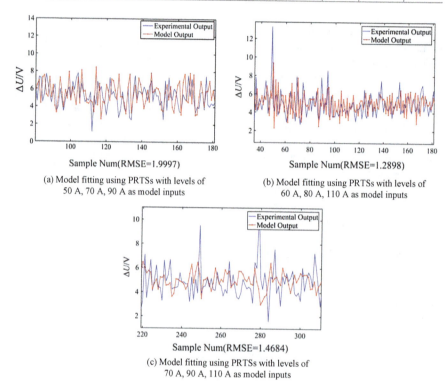

Fig. 9 System identification experimental results of different levels of PRTSs

identified model under different PRTSs as system input using the model parameters in Table 4. Figure 9a–c are identification results for model parameters set number 1, 2 and 3 in Table 4. From Fig. 9, the model output and actual system output were in an agreement, and the calculated RMSE was within an acceptable range. The value of RMSE in Fig. 9a was the largest one compared to other two groups in Fig. 9, but the model output can also reflect the variation of actual system output well. The model parameters set number 1 in Table 4 corresponding to Fig. 9a was selected to validate the accuracy of the Hammerstein model.

$$\text{RMSE} = \frac{\frac{1}{N}\sum_{k=1}^{N}[y(k) - \hat{y}(k)]^2}{\frac{1}{N}\sum_{k=1}^{N}[y(k) - \overline{y}(k)]^2} \tag{13}$$

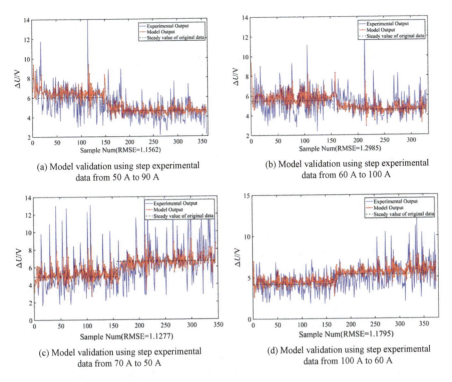

Fig. 10 Hammerstein model with disturbances

where y represents the experimental actual outputs, $\hat{y}$ represents the model fitting output, and $\bar{y}$ represents mean value of experimental outputs.

According to the identified model parameters set number 1 in Table 4, the Hammerstein model can be described as Eq. (14). Figure 10 showed the results of model validation using different step experiments data. Analyzing the fitting results, the Hammerstein model with disturbances fitted very well with the actual system, and the model output can reach the steady-state value and fluctuate with time near it same as actual system. And the RMSE was in an acceptable range.

In a comprehensive view, the weld penetration control system is a time-varying nonlinear system with time delay and can be described by Hammerstein model with disturbances, which was composed of nonlinear static system and linear dynamic system. The complicated control system of weld penetration was simplified to classical system, which is advantageous for system control. However, the influence factors of system are quite a lot and unpredictable in GMAW-P process, thus, it is necessary to add adaptive control algorithm to the control system.

$$\begin{cases} y(k) = 0.2686 * y(k-1) - 0.2795 * y(k-2) - 0.1618 * y(k-3) \\ \quad -0.6375 * x(k-d) - 1.004 * x(k-d-1) + 1.1231 * x(k-d-2) \\ \quad +1.1248 * x(k-d-3) + 0.2685 * x(k-d-4) + 0.3102 * v(k-1) \end{cases}$$
(14)

where $x(k) = \frac{1}{1.4347}(0.00076 \times u^2(k) - 0.155 \times u(k) + 12.5)$.

4 Conclusion

The change in arc voltage during peak current period (ΔU) was used to control the depth of weld penetration in GMAW-P. A nonlinear system model was established to control the weld penetration with base current (I_b) as system input and ΔU as system output. The nonlinear control system was identified by a proposed nonlinear Hammerstein model algorithm and some conclusions were obtained as followings.

1. Base current (I_b) was selected as the system input, and the step response experiments were conducted to get system characteristics. The relationship between I_b and ΔU was analyzed in steady state, and a nonlinear relation was established between I_b and ΔU.
2. A Hammerstein model with disturbances was used to describe the nonlinear penetration control system, which consisted of nonlinear static system and linear dynamic system. The polynomial relation between I_b and ΔU in steady state was selected as the function of nonlinear static system with a coefficient, which was the gain for linear dynamic system. The step response experiments data were selected to validate the accuracy of identified model. The model output closely conforms to the actual system output within an acceptable range error. It is suitable to describe the weld penetration using a nonlinear Hammerstein model with disturbances.
3. The model parameters were obtained with different PRTSs as inputs, by the recursive least square identification algorithm. But the identified system model parameters are time-varying. Thus, an adaptive identification algorithm should be added to the control process.

Acknowledgements This research is supported by the National Natural Science Foundation of China (51505326), the Natural Science Foundation of Tianjin (16JCQNJC04300).

References

1. Wang ZJ, Zhang YM, Wu L (2010) Measurement and estimation of weld pool surface depth and weld penetration in pulsed gas metal arc welding. Weld J 89(6):117s–126s
2. Pal K, Pal SK (2011) Effect of pulse parameters on weld quality in pulsed gas metal arc welding: a review. J Mater Eng Perform 20(6):918–931

3. Saeed G, Zhang YM (2007) Weld pool surface depth measurement using a calibrated camera and structured light. Meas Sci Technol 18(8):2570
4. Xiao YH (1993) Weld pool oscillation during GTA welding of mild steel. Weld J 72(8):428s–434s
5. Nagarajan S, Banerjee P, Chen W et al (1992) Control of the welding process using infrared sensors. IEEE Trans Robot Autom 8(1):86–93
6. Fan C, Lv F, Chen S (2009) Visual sensing and penetration control in aluminum alloy pulsed GTA welding. Int J Adv Manuf Technol 42(1–2):126–137
7. Ramos EG, Caribé Guilherme, de Carvalho Sadek, Alfaro Crisóstomo Absi (2015) Analysis of weld pool oscillation in GMAW-P by means of shadow graphy image processing. Weld Int 29(3):9
8. Wang Z, Zhang YM, Wu L (2012) Adaptive interval model control of weld pool surface in pulsed gas metal arc welding. Automatica 48(1):233–238
9. Wang Z, Zhang YM, Wu L (2011) Predictive control of weld penetration in pulsed gas metal arc welding. Robot Weld Intell Autom 2011:263–269
10. Liu YK, Chen SJ, Zhang WJ et al (2013) Nonlinear dynamic modelling of weld penetration in gas tungsten arc welding process. Adv Mater Res 658:292–297
11. Liu YK, Zhang YM (2013) Weld penetration control in gas tungsten arc welding (GTAW) process. In: IECON 2013—39th annual conference of the IEEE industrial electronics society, IEEE, pp 3842–3847
12. Liu Y (2013) Estimation of weld joint penetration under varying GTA pools. Weld J 92(11):313s–321s
13. Ding F, Chen T (2005) Identification of Hammerstein nonlinear ARMAX systems. Pergamon Press
14. Narendra K, Gallman P (2003) An iterative method for the identification of nonlinear systems using a Hammerstein model. IEEE Trans Autom Control 11(3):546–550
15. Lang ZQ (1997) A nonparametric polynomial identification algorithm for the Hammerstein system. IEEE Trans Autom Control 42(10):1435–1441
16. Chang F, Luus R (1971) A noniterative method for identification using Hammerstein model. IEEE Trans Autom Control 16(5):464–468
17. Bai EW (1998) An optimal two stage identification algorithm for Hammerstein-Wiener nonlinear systems. In: Proceedings of the 1998 American control conference, ACC, pp 2756–2760
18. Na X, Zhang YM, Liu YS et al (2010) Nonlinear identification of laser welding process. IEEE Trans Control Syst Technol 18(4):927–934
19. Ye X, Hu L, Liu Y (2009) Nonlinear identification and self-learning CMAC neural network based control system of laser welding process. In: 9th international conference on electronic measurement & instruments, IEEE, pp 3440–3445
20. Wang Z (2010) Adaptive interval model control for depth of weld penetration in pulsed gas metal arc welding. Dissertation, Harbin Institute of Technology

Wandong Wang is currently working at Huawei Investment & Holding Co., Ltd. He received his B.S. degree in Material Processing and Control Engineering (Welding Major) from Nanjin University of Science and Technology in 2016 and his M.S. degree in Material Process Engineering (Welding Major) from Tianjin University in 2019. His research interests include welding process control and innovative welding processes.

Zhijiang Wang is currently an associate professor in Tianjin University, where he received his B.S. degree in Metal Material Engineering (Welding Major) in 2003 and has worked since 2010. He received his Ph.D. and M.S. degree in Material Process Engineering (Welding Major) from the State Key Laboratory of Advanced Welding Production Technology at the Harbin Institute of Technology, China. Dr. Wang, as a visiting scholar, worked with Dr. Yuming ZHANG at the University of Kentucky for more than two years since 2008. His research interests include innovative welding processes, monitoring and control of welding processes and welding automation.

# Effect of Transverse Ultrasonic Vibration on MIG Welded Joint Microstructure and Microhardness of Galvanized Steel Sheet

Guohong Ma, Xiaokang Yu, Jian Li and Yinshui He

Abstract A comparison test of conventional MIG welding and ultrasonic-MIG hybrid welding was carried out in this paper. The effects of transverse ultrasonic vibration on weld formation, weld microhardness and weld microstructures during ultrasonic-MIG hybrid welding of 1 mm thick galvanized steel sheet were discussed. Microstructures of weld were analyzed with optical microscopy and scanning electron microscopy, and microhardness of weld joint was measured with Vickers hardness tester. The results show that the grains in welded zone of ultrasonic-MIG hybrid welding are finer and uniformly distributed; the hardness of the whole weld zone is more uniform; weld width increases; and weld depth and residual height decrease compared with the conventional MIG welding.

Keywords Galvanized steel sheet · Ultrasonic-MIG · Microstructure · Microhardness

1 Introduction

Galvanized steel sheet has good corrosion resistance. The main reason is that zinc is a kind of amphoteric metal, which is easy to be corroded by acid or strong alkali. In neutral or weakly alkaline medium, Zinc oxide, zinc hydroxide and zinc carbonate formed on the surface of zinc will hinder continued corrosion of the steel plate [1]. In different corrosion environments, common steel plates are corroded 25 times faster than galvanized steel plates [2]. Because of the better corrosion resistance and good

---

G. Ma · X. Yu · J. Li
School of Mechanical Engineering, Key Laboratory of Lightweight and High Strength Structural Materials, Nanchang University, Xuefu Road 999, Nanchang 330031, Jiangxi, China
e-mail: my_126_my@126.com

Y. He (✉)
School of Resource Environmental and Chemical Engineering, Nanchang University, Xuefu Road 999, Nanchang 330031, Jiangxi, China
e-mail: heyingshui117@163.com

© Springer Nature Singapore Pte Ltd. 2019
S. Chen et al. (eds.), *Transactions on Intelligent Welding Manufacturing*, Transactions on Intelligent Welding Manufacturing, https://doi.org/10.1007/978-981-13-7418-0_7

plasticity of galvanized steel, it has been widely used in engineering. In particular, it has a great application prospect in automotive light weighting and improving service life [3]. According to statistics, within 10 years of automobile use, automobile failure rate produced by galvanized steel sheet is about 20% lower than that produced by non-galvanized steel sheet [4]. Wide application of galvanized steel sheets makes welding of galvanized steel sheets more important. At present, the welding methods commonly used for galvanized steel sheets are spot welding and laser welding. When resistance spot welding is used, the galvanized layer easily alloys the electrode with the zinc layer, reducing the life of the electrode [5]. Laser welding has the advantages of small deformation of weldment, high-efficiency welding and easy automation. However, due to the high energy density of the laser, the weld cooling speed is fast. The gas generated by zinc gasification during welding of galvanized steel is not easy to escape from the weld [6]. MIG welding has the characteristics of high-efficiency welding, stable welding quality and all-position welding. It has been widely used in actual production [7]. With the continuous development of welding technology, more and more attention has been paid to the hybrid welding method combining two or more heat sources. Hybrid welding is a new high-efficiency welding method that combines the advantages of a variety of single welding methods to form a more advanced welding method [8]. The hybrid welding method that utilizes the unique mechanical properties of ultrasonic wave, and the high energy density to combine with welding arc has been widely concerned.

In 2002, Dai Wenlong of Taiwan Feng-Chia University have studied the ultrasonic vibration head vertically acts the upper surface of base metal in the welding process of aluminium alloy 7075-T6 [9]. The study found that the ultrasonic field can significantly change the microstructure of the weld heat affected zone and the weld zone. The grain size of the weld heat affected zone is reduced, and the weld depth is increased by about 45%. In 2009, Watanabe et al. of Niigata University of Japan used gas metal arc welding to weld stainless steel, and ultrasonic waves were introduced into the weld pool by welding wire [10]. It has been found that at lower welding speeds, the grains are significantly refined; at higher welding speeds, many equiaxed grains are formed in the center of the weld perpendicular to the welding direction, and the presence of equiaxed grains leads to elongation of the welded joint is increased by about 40%. Harbin Institute of Technology Yang Chunli et al. proposed a hybrid welding method that directly acted ultrasonic waves to the arc and the molten pool [11, 12], which is to act the ultrasonic vibration longitudinally to the TIG welding process. Through the conventional TIG and ultrasonic-TIG plate surfacing experiments on 2014 aluminum alloy, the results show that the ultrasonic-TIG welding not only improves the penetration depth and depth-to-width ratio of the weld but also refines the grains and change crystallization form under the action of ultrasonic waves [13]. In this paper, the conventional MIG welding method and the ultrasonic-MIG hybrid welding method are used to carry out the plate surfacing test on the galvanized steel sheet. The influence of transverse ultrasonic vibration on the weld formation and joint microstructure of galvanized steel sheet MIG welding was discussed.

2 Test Materials and Methods

The base metal is Q235 galvanized steel sheet, its size is 250 mm × 50 mm × 1 mm. The welding wire is a normal carbon steel wire with a diameter of 1.2 mm. The chemical composition of the base metal and the welding wire are shown in Tables 1 and 2, respectively.

The test was used ordinary MIG welding and ultrasonic-MIG hybrid welding to conduct plate surfacing welding. The principle of ultrasonic-MIG welding is to apply ultrasonic vibration laterally to the end of the MIG welding torch.

Figure 1 is a diagram of the ultrasonic-MIG hybrid welding system. Ultrasonic horn transverse contacts the end of MIG welding torch nozzle. The ultrasonic transducer converts the high-frequency electrical signal provided by the ultrasonic generator into mechanical vibration through the piezoelectric effect of the material. It is transmitted to the welding gun through the horn and acts on the end of the welding wire. The distance from the end of the protection nozzle to the workpiece is 1.2 mm. The welding power source of this test is Huayuan NB-350 IGBT type MIG welding machine, and the ultrasonic equipment type is ZJS-500 ultrasonic generator. High purity argon with a mass fraction of 99.9% is used as a shielding gas in the welding. Welding parameters: welding current 50 A, welding voltage 23.4 V, welding speed 6.2 mm/s, and gas flow 15 L/min. Ultrasonic equipment parameters: frequency 28 kHz, working voltage 220 V, working current 0.6 A, and output power 300 W. Sampling the weld cross section of galvanized steel sheet after welding to analyze weld formation. The parameters such as weld width, weld depth and residual height are measured by a weld inspection ruler. The specimens were inlaid, ground, polished, and corroded according to standard metallographic procedures, and then, the weld microstructure was analyzed using a BX51M optical microscope and an S-3000N scanning electron microscope. At the same time, the microhardness of the base metal, fusion zone and weld zone of the sample was measured by the HXS-1000A digital intelligent Vickers microhardness tester.

Table 1 Chemical composition of galvanized steel sheet

Composition	C	Mn	P	S	Fe
ω/%	≤0.15	≤0.6	≤0.05	≤0.05	Bal

Table 2 Chemical composition of welding wire

Composition	C	Mn	Si	P	S	Cu	Cr	Ni	Mo	V
ω/%	0.06–0.15	1.40–1.85	0.08–1.15	0.025	0.025	0.05	0.15	0.15	0.15	0.03

Fig. 1 Ultrasonic-MIG hybrid welding system apparatus

3 Results and Discussion

3.1 Effect of Transverse Ultrasonic Vibration on Welded Formation

The macroscopic morphology and cross section of the weld obtained by welding the galvanized steel sheet by ultrasonic-MIG welding and conventional MIG welding method are shown in Fig. 2. It can be seen that the weld-forming condition is good, and there are no obvious macro-defects. However, through comparison, it can be found that the ultrasonic-MIG hybrid welding has better weld formation, of which the distribution of fish scale pattern is uniform, and the weld width is stable. The values of weld width, penetration and reinforcement were measured with a weld inspection ruler. In order to make its results reasonable, the average value were calculated based on the measurement results of three positions on a weld.

Figure 3 is a comparison of the weld size data of ultrasonic-MIG welding and conventional MIG welding. It can be clearly seen from the figure that the MIG

Effect of Transverse Ultrasonic Vibration on MIG ...

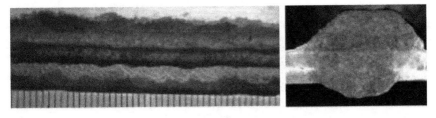

(a) Conventional MIG welding

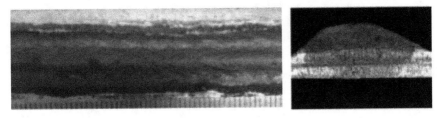

(b) Ultrasonic-MIG hybrid welding

Fig. 2 Image of appearances and cross section of welded seams

welding seam to which the transverse ultrasonic vibration is applied has a wider width, the weld depth and the residual height smaller. The size of weld width of conventional MIG welding varies greatly and unstable, while the size of ultrasonic-MIG welding seam is relatively uniform.

Since the ultrasonic-MIG hybrid welding and conventional MIG welding are performed under the same welding parameters, the total heat input to the weldment is constant. By applying transverse ultrasonic vibration, the heat distribution in the direction of vertical welding seam is wider, which is equivalent to reducing the heat at a certain point of weldment, so that the penetration depth of the weld is reduced and the weld width is increased. At the same time, the transverse ultrasonic vibration also makes the droplet transition more stable, so the weld morphology is more uniform and the weld width is more stable.

3.2 Effect of Transverse Ultrasonic Vibration on Welded Joints Microhardness

Figure 4 shows the microhardness values of different positions in the conventional MIG welding and ultrasonic-MIG hybrid welding. It can be seen that the hardness values of each point in the weld zone of ultrasonic-MIG hybrid welding are not much different, while conventional MIG welding. The hardness value fluctuates greatly, with a maximum hardness of 354.6 and a minimum hardness of only 194.3. Therefore,

Fig. 3 Effect of transverse ultrasonic vibration on welded sizes

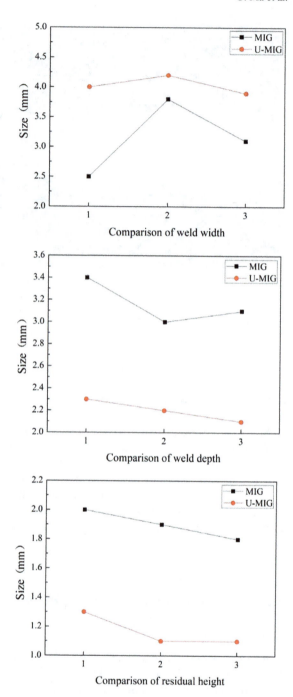

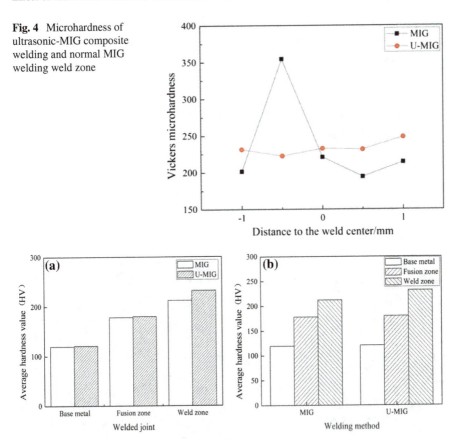

Fig. 4 Microhardness of ultrasonic-MIG composite welding and normal MIG welding weld zone

Fig. 5 Average hardness value of welded joints

the hardness of the entire weld zone can be more uniform by applying transverse ultrasonic vibration.

In order to reduce the error of hardness measurement, removing the maximum and minimum hardness values in the same area to obtain the average hardness values in different areas of the welding joint, as shown in Fig. 5. It can be seen from Fig. 5 that the hardness of the base metal is low, the hardness value is only 120, and the average hardness of the weld zone is the highest. This is because the ambient temperature during welding is low and the weld is cooled quickly, which is equivalent to quenching treatment. The average hardness value of the weld zone is always greater than that of the base metal, whether it is ordinary MIG welding or ultrasonic-MIG welding. Figure 5b shows the comparison of the hardness values of conventional MIG welding and ultrasonic-MIG welding in the same area. It can be seen that the average hardness value of the ultrasonic-MIG hybrid welding weld zone is significantly higher than that of conventional MIG welding. The main reason is that the lateral ultrasonic vibration makes the microstructure of the weld zone finer, the distribution is more uniform, so as to improve the hardness of the weld. Therefore, by applying transverse ultrasonic vibration, the hardness of the MIG welding seam region of the galvanized steel sheet can be improved and the hardness value can be made more uniform.

3.3 Effect of Transverse Ultrasonic Vibration on Welded Seams Microstructural

In order to study the effect of transverse ultrasonic vibration on the microstructure of weld seams of galvanized steel plate, the central area of weld seams was photographed by metallographic microscope and scanning electron microscope, as shown in Fig. 6.

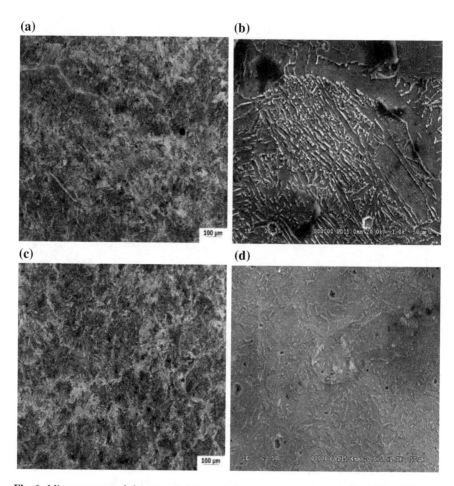

Fig. 6 Microstructure of the center of the welded seams: **a, b** conventional MIG welding; **c, d** ultrasonic-MIG hybrid welding

Figure 6a, c are metallographic microscope photographs, while Fig. 6b, d are SEM photographs. It can be seen from Fig. 6a, c that the microstructure of the weld area of galvanized steel plate is mainly composed of ferrite and pearlite. It can also be found that the distribution of ferrite and pearlite in the weld is more uniform under the action of transverse ultrasonic vibration, otherwise there are coarse ferrite and pearlite in the weld. It can be clearly seen from Fig. 6b, d that some of the ferrite grains are large in the conventional MIG welding, some are small, and the pearlite distribution is also uneven. And the grain distribution in the weld zone is uniform through the ultrasonic-MIG hybrid welding process, and its grain size is finer. This is due to the large range of thermal input in the process of applying transverse ultrasonic vibration in MIG welding, which makes the average thermal input of weldment becomes smaller, resulting in austenite grain growth is not obvious, and the ferrite after cooling is finer.

4 Conclusion

The transverse ultrasonic vibration will affect the weld formation of the galvanized steel plate MIG welding, after ultrasonic vibration is applied, weld width increases and weld depth and residual height decreases.

The microstructure of the weld is composed of ferrite and pearlite. After transverse ultrasonic vibration is applied, the ferrite distribution in the weld is more uniform and the grains are finer.

The microhardness value of the weld zone to which the transverse ultrasonic vibration-assisted welding is applied is more uniform, and the average hardness value is superior to the MIG weld to which no ultrasonic vibration is applied.

Acknowledgements This research was supported by the National Natural Science Foundation of China (51665037) and the Key Laboratory of Lightweight and High Strength Structural Materials of Jiangxi Province (20171BCD40003).

References

1. Zhan WH, Liu HJ, Cao C et al (2010) Pretreatment process of electroless nickel plating on mould zinc alloy surface. Corros Sci Prot Technol 22(3):220–223
2. Yu JS, Zhang JX, Wu JS et al (2005) Review of properties of hot-dip galvanized steel coatings for automobiles. Phys Chem Test Phys 41(7):325–328
3. Tan J, Wang J, Gao HY et al (2008) Research progress of high strength steel alloy hot dip galvanizing. Mater Rev 22(2):64–67
4. Liu CD, He GF, Chen HY (2006) Study on weldability of galvanized steel sheets for automobile covers. J Ordnance Equip Eng 27(1):38–40
5. Dasgupta AK, Mazumder J (2008) Laser welding of zinc coated steel: An alternative to resistance spot welding. Sci Technol Weld Joining 13(3):289–293

6. Kim JD, Na H, Park CC (1998) CO_2 laser welding of zinc-coated steel sheets. KSME Int J 12(4):606–614
7. Yang X (2012) Research on arc characteristics and weld line formation mechanism during TIG welding controlled by electromagnetic fields. Dissertation, Shenyang University of Technology
8. Chen YB (2005) Modern laser welding technology. Science Press, Beijing, p p103
9. Dai WL (2003) Effects of high-intensity ultrasonic-wave emission on the weldability of aluminum alloy 7075-T6. Mater Lett 57(16):2447–2454
10. Watanabe T, Shiroki M, Yanagisawa A et al (2010) Improvement of mechanical properties of ferritic stainless-steel weld metal by ultrasonic vibration. J Mater Process Tech 210(12):1646–1651
11. Sun QJ, Lin SB, Yan CL et al (2008) The arc characteristic of ultrasonic assisted TIG welding. China Weld 17(4):65–69
12. Sun QJ, Lin SB, Yang CL et al (2010) Development and application of ultrasonic-TIG hybrid welding device. Trans China Weld Inst 31(2):79–82
13. Yuan H, Lin S, Yang C et al (2011) Microstructure and porosity analysis in ultrasonic assisted TIG welding of 2014 aluminum alloy. China Weld 20(1):39–43 (English version)

Guohong Ma Professor, Doctoral tutor, Research direction: Intelligent welding, Advanced welding technology. He got Ph.D. degree in 2002.03–2006.03 at Shanghai Jiao Tong University.

Research field Material Processing Engineering (Major): The main research areas are intelligent welding technology and advanced welding methods. Research on visual sensing, collaborative control, information transmission and modeling based on welding robot is carried out. Research on DE-GMAW welding method and MIG hybrid welding method was carried out.

Yinshui He Research direction: Intelligent control, Welding robot. He got Ph.D. degree in 2012.09–2017.03 at Shanghai Jiao Tong University.

Research field Control Science and Engineering (Major): The main research area is intelligent technologies for robotic welding, including vision sensor technologies, pattern recognition technologies and weld seam tracking and control. The specific research contains visual attention modeling, clustering, neural networks, hybrid system modeling and model predictive control, etc.

Effect of Scanning Mode on Microstructure and Physical Property of Copper Joint Fabricated by Electron Beam Welding

Ziyang Zhang, Shanlin Wang, Jijun Xin, Yuhua Chen and Yongde Huang

Abstract Copper T2 thin sheet thickness of 2 mm by vacuum electron beam welding, scanning electron microscopy, optical microscopy, microhardness tester, and tensile testing machine was used. The formation of the weld surface by different scanning methods was studied. The influence of cross section morphology and microhardness of welded joints and The Effect of microstructure on mechanical properties of the welded joint. The results showed that in working distance is 300 mm and the accelerating voltage $U = 60$ kV, focusing current $I_f = 502$ mA, welding speed $V = 800$ mm/min and electron beam flow $I_b = 27$ mA situation, without the addition of scanning, the welded joint with good forming and no macroscopic defects can be obtained. Under the same conditions contrast to add the scanning mode of welding and derived welding in the process of adding triangle wave scanning, not only the forming good welded joints can be attained, but also the welding joints hardness values and tensile strength are improved. The average tensile strength of the welded joint is 178.8 MPa, the maximum tensile strength can reach 85% of the base metal, and the elongation rate is 84.5%, which is the weakest in the weld fusion zone compared with the whole weld zone.

Keywords T2 copper · Electron beam welding · Weld formation · Mechanical property

---

Z. Zhang · S. Wang (✉) · Y. Chen · Y. Huang
National Defence Key Disciplines Laboratory of Light Alloy Processing Science and Technology, Nanchang Hangkong University, Nanchang 330063, Jiangxi, China
e-mail: slwang70518@nchu.edu.cn

Z. Zhang
e-mail: m15270808960@163.com

Z. Zhang
Nanjing Engineering Institute of Aircraft Systems, Nanjing 211100, Jiangsu, China

J. Xin
Institute of Plasma Physics, Chinese Academy of Sciences, Hefei 230031, China

© Springer Nature Singapore Pte Ltd. 2019
S. Chen et al. (eds.), *Transactions on Intelligent Welding Manufacturing*,
Transactions on Intelligent Welding Manufacturing,
https://doi.org/10.1007/978-981-13-7418-0_8

1 Introduction

Due to their high electrical conductivity, thermal conductivity, ductility, and relatively low production cost, the copper alloy and their composites have been applied extensively in many industrial fields, such as petroleum, chemical, marine communications, and power systems [1–4]. In many cases, the various combinations of positions must be connected together, so the weld can be achieved [5]. The welding method of copper T2 (commonly called industrially pure copper) includes generally shielded metal arc welding, TIG, and brazing. Shielded metal arc welding must be used electrode increasing the materials cost, moreover, manual arc welding should be in a well-ventilated environment preventing copper poisoning [6]. Though the high energy for the laser welding, two major problems are difficult to overcome. One is the copper and copper alloy with high reflectivity and conductivity, the other is weld defects of copper and copper alloy welded with laser welding, such as porosity and splash [7–12]. Electron beam welding technology has the advantages of high energy density, small heat input, precise control of energy input, small welding deformation and vacuum welding without air pollution. It also has unique advantages in the connection of materials with higher thermal conductivity [13].

But there are few types of research on copper T2 vacuum electron beam welding. Therefore, this article mainly by changing welding in the process of scanning waveform, of T2 copper of vacuum electron beam welding test, study the different scanning waveforms on the welding joint microstructure and mechanical properties, in order to obtain high quality of T2 copper welding joints to provide the experimental basis.

2 Experimental

The copper T2 plate with the size of 70 mm × 40 mm × 2 mm is welded. Before welding, the weld surface should be cleaned with iron-grinding brush and acetone. The copper joint is fabricated using KS15-PN150KM electronic beam welder, with the weld parameters of the working distance D 300 mm, focusing surface current for 502 mA, accelerating voltage U 60 kV, welding speed V 800 mm/min and electron beam flow I_b 27 mA, scanning methods: non-scanning, sawtooth wave scanning, elliptical scanning, circular scanning, or triangle wave scanning as shown in Table 1. After welding, the use of wire cutting in the cross section of the weld was separately cut metallographic and tensile test specimens, and Fig. 1 is tensile specimen size. The metallographic sample is made of $FeCl_3$ 10 g + HCl 6 mL + C_2H_5OH 20 mL + H_2O 80 mL, which is composed of the reagent for corrosion, the microstructure was observed by XJP-2C optical microscope. The tensile strength of the joint was tested on the WA-100 electronic universal material testing machine according to the GB/T2651-2008, and the tensile rate is 2 mm/min, taking the average value of the three samples, and the fracture morphology is analyzed by SEM after drawing. The

Effect of Scanning Mode on Microstructure …

Table 1 Under different scanning methods process parameters

Scanning methods	Process parameters	Scanning amplitude (mm)	Frequency (Hz)	Waveform
Non-scanning	V = 800 mm/min	/	/	/
Sawtooth wave scanning	I_b = 27 mA	x = y = 0.5		⁄\
Elliptical scanning	U = 60 kV	x = 0.5, y = 0.2	500	○
Circular scanning	D = 300 mm	x = 0.6, y = 0.4		○
Triangle wave scanning	I_f = 502 mA	x = 0.6, y = 0.5		∧

Fig. 1 Tensile specimen

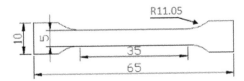

Fig. 2 Resistance sample

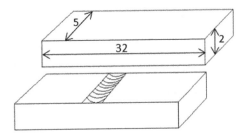

hardness of the joint was measured by the 401MVD digital microhardness tester, the load was 100 N, and the loading time was 5 s. The sample is cut perpendicularly along the weld center by wire cutter and then the cutting surface should be polished (Fig. 2). At normal temperature, the resistance was measured by QJ36S-2 DC low resistance tester.

3 Results and Discussion

The macroscopic morphology of the joints fabricated with different scanning waveforms is shown in Table 2. Obviously, when there is no scanning waveform, weld undercut and collapse phenomenon, the weld formation is poor, as shown in Table 2(a). The surface forming of the welded joint is increased when the surface of the welded joint is formed when the scanning is carried out, but the surface mor-

Table 2 Under different scanning methods process image

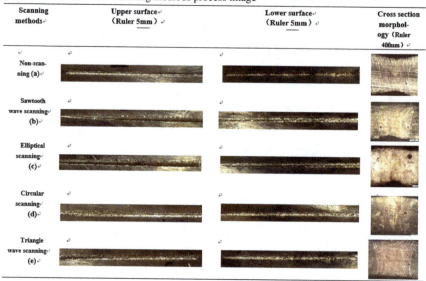

phology of the welded joint is better than that of the non-scanning. A "solidification streamline" can be observed from the cross-section of joint for all sample, and the "solidification streamline" seems distributed s symmetrically alone the weld center. The "solidification streamline" is dense at the top layer of weld while sparse at the bottom layer of weld. Some pores are formed along the fusion line in all samples. Due to the high thermal conductivity of copper, a rapid cool rate and a residence time at high temperature can be deduced from the "solidification streamline". Compared with the cross-sectional morphology of the weld, the weld penetration state of the triangle wave scanning method is obtained. To a certain extent, the method of triangle wave scanning is of high energy utilization rate compared to other scanning methods, and the uniformity of the distribution of energy density is better.

The main defects are the pores for the copper joint fabricated by electron beam welding as shown in Table 2. And, a unique "solidification streamline" organization has appeared in the welded joint.

Due to the similar streamline appearance for all samples, a typical morphology of joint welded with triangle wave scanning is shown in Fig. 3. In liquid metal, along with the crystal growth direction, if the temperature gradient exceeds the critical value of 10 °C/cm [14], convection will make the front of solid–liquid interface temperature fluctuation, which makes the interface to promote the speed of the onset of the disorder. When the interfacial temperature increases, the interfacial advancing velocity decreases with the decrease of the concentration of solid phase in the solid phase. On the contrary, when the interface temperature is lower, the velocity of the interface is increased, the solute concentration in the solid phase is increased, resulting in the formation of zonal segregation in the solid phase. Due to the horizontal

Effect of Scanning Mode on Microstructure ...

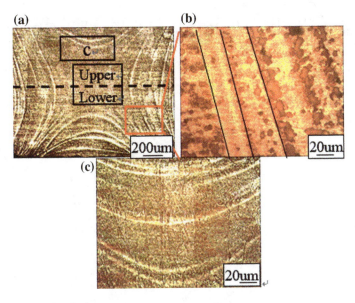

Fig. 3 Streamline distribution

solidification shrinkage and the vertical gravity causes by the streamline, the molten metal is flowing from the right of top weld layer to the left of below weld layer, which is driven by the comprehensive effects of the two forces. It is well known that the isotherms and directions opposite to the movement of molten metal, so negative segregation is occurred. In the bottom part of joint, the gravity is mainly force, and liquid flow direction and the direction of the isotherm is consistent, so the positive segregation is occurred. In this case, the liquid flow is caused by the solidification shrinkage, and the flow direction of the liquid is opposite to that of the isotherm. It can also be concluded that the upper part of the weld pool is more severe than the lower part of the upper part of the weld pool by Fig. 2, melt diffusion is also more uniform, and shows that the liquid flow rate is also smaller farther away from the fusion line of the molten pool.

Contrast to the non-scanning, circular scanning, triangle wave scanning, some defects were occurred near the fusion line area and the upper region of weld center with the scanning of sawtooth wave scanning and elliptical scanning. As shown in the fine grains at the grain boundaries of coarse grains, showing a typical "necklace" structure, it is deduced that the dynamic recrystallization occurred during solidification. With the increase of deformation temperature, recrystallization volume percent increased significantly. Coarse original grains are replaced by fine recrystallization grain. When the temperature continues to increase, the crystalline grains will grow. For the copper joint welded by electron beam welding, the heat-affected zone is very narrow, and a fine grain zone is formed near the fusion zone, as shown Fig. 4a which is welded without the scanning mode. Figure 4b circular scanning and Fig. 4c of circular scanning mode of weld nearly fusion line microstructure in fine grain zone

Fig. 4 Near fusion line and the microstructure of the upper part of the weld: **a** non-scanning, **b** circular scanning, **c** triangle wave scanning

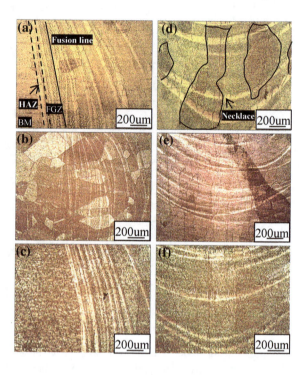

and the columnar crystal appear alternately, i.e., the temperature stratification, indicating that the two scanning modes weld microstructure appeared a certain degree of segregation. Also by Fig. 4d–f, it can be clearly seen in the upper part of the weld center due to the heat is more concentrated and coarse grain. For grain coarsening, most of the copper and copper alloys in the welding process, the general does not occur solid phase transformation, the weld is a crystal of coarse columnar crystal. For grain coarsening, most of the copper and copper alloys in the welding process, the general does not occur in solid-state phase transition. And copper alloy weld metal grain growth, also make the mechanical properties of the joint reduction. According to the traditional theory of the solidification for weld metal, the columnar crystal structure is formed along the weld pool edge and grown with epitaxial growth, and the solidification of the weld metal crystal is observed universally [15].

Figure 5 the microhardness distribution along the central line of the cross-section of the welded joint with no scanning, triangle wave scanning and circular scanning. It can be seen the microhardness of pure copper joint is symmetrical distribution to weld centerline. The lowest Vickers hardness of $H_{V0.1}$ 42–45 is occurred in weld zone. From the weld center, hardness value increased gradually, and near the HAZ, the hardness value decreased gradually. It is can be deduced that electron beam welding will make copper welding joints of a certain degree of softening. In contrast the microhardness curve, non-scanning triangle wave scanning, circular scanning, can see the triangle wave weld microhardness of the region of the average value is higher than that of circular scanning, mean values of circular scanning microhardness

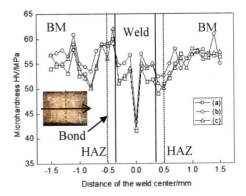

Fig. 5 Microhardness curves of joints under different scanning modes: a non-scanning, b triangle wave scanning, c circular scanning

is higher than that of the non-scanning, and the microhardness of circular scanning value in weld area relatively large fluctuations, and the average hardness values is higher than that of no scanning, and lower than that of triangle wave scanning.

The microhardness of the triangular wave scanning method is higher than that of the other samples, mainly due to the small size of the triangular wave scanning method. According to Holzer Paige formula [16], i.e.,

$$H_V = H_0 + k_H d^{-1/2} \tag{1}$$

where H_V for microhardness; d for grain diameter; H_0 and k_H are appropriate constants associated with hardness measurements. It can be concluded that the microhardness of the triangular wave scanning method is higher than other scanning methods, and it is caused by the fine grain size of the triangular wave scanning method.

As shown in Fig. 6, (a) non-scanning, (b) triangle wave scanning, (c) circular scanning, (d) base metal sample by analyzing the macroscopic tensile fracture morphology of the welded joint, it is obvious that the necking of the base metal is more obvious, that is, the toughness of the base metal is better than the toughness of the welded joint. Due to a stress concentration near the fusion line, the deformation gradually increased and the macrocrack was initiated when the stress is greater than

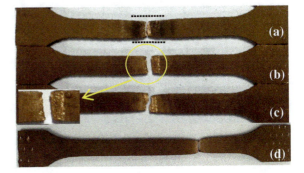

Fig. 6 Macroscopic tensile specimen: a non-scanning, b circular scanning, c triangle wave scanning, d base metal

the weld-metal-bearing capacity. The fracture is located in the fusion zone, so the weld fusion zone is the weak part of the joint.

Tensile test data is shown in Fig. 7. It can be seen from Fig. 7 that when the scanning mode changes from a circular scan to triangular wave scanning, the trend of the tensile curve is approximately the same. Moreover, it can be seen from the tensile curve that the tensile strength and elongation after welding are weaker than that of base metal. From the Table 3 that when the scanning mode is changed from circular scanning to triangular wave scanning, the tensile strength of the welded joint shows an increasing trend, but it is lower than the tensile strength of the base metal. For the extension rate, when the scanning mode changes from circular scanning to triangular wave scanning, the trend of the elongation of the welding joint is the same as that of the tensile strength. The results show that the average tensile strength of the electron beam welded joint is 178.8 MPa, and the tensile strength of the tensile joint can reach 85% of the base metal, and the elongation is up to about 84.5% of the base material elongation.

The conductivity can be measured according to GB standard volume resistivity, as follows:

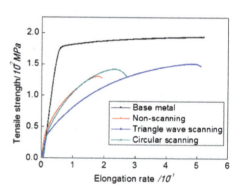

Fig. 7 Tensile curve

Table 3 Tensile test data under different scanning waveforms

Scanning methods	Circular scanning	Non-scanning	Triangle wave scanning	Base metal
Tensile strength (MPa)/elongation rate	$a = 178/a = 0.152$	$a = 163.6/a = 0.086$	$a = 190.2/a = 0.168$	$a = 219/a = 0.2253$
	$b = 176.5/b = 0.172$	$b = 178.1/b = 0.103$	$b = 169.7/b = 0.185$	$b = 220.47/b = 0.205$
	$c = 175.8/c = 0.149$	$c = 189.7/c = 0.157$	$c = 199/c = 0.193$	$c = 221.36/c = 0.215$
Average tensile strength (MPa)	176.77	177.33	186.4	220.26
Average elongation	0.107	0.1153	0.182	0.2153

$$\rho_{20} = R_{20}(A_{20}/L_{20}) \tag{2}$$

ρ_{20} Volume resistivity at 20 °C, Ω mm²/m;
R_{20} Sample resistance value at 20 °C, Ω;
A_{20} Specimen cross-sectional area at 20 °C, mm²;
L_{20} Sample measurement length at 20 °C, m;

Resistivity results are shown in Fig. 8. It can be concluded that the weld has an effect on the resistivity of copper alloy. The circular scanning waveform has the greatest influence on the weld resistivity, and the triangular scanning waveform has the least influence on the resistivity. Due to defects as the gas pore, heat crack, some alloying elements dissolved, the heterogeneous atoms, dislocations, and impurity elements in Cu crystals, the conductivity of copper joint was decreased [17]. It has been reported that imperfections such as grain boundaries, dislocations, or impurity atoms serve as scattering centers for conduction electrons in metals and decreasing in the number of these imperfections raises the electrical conductivity [18, 19]. For example, alloying pure copper may increase the strength by two or three times, but the electrical conductivity of copper alloys is only 10–40% that of pure copper [18]. After the thermal cycle of the electron beam welding process, the grain boundary area and the dislocation density in both HAZ and FZ would significantly decrease [20]. The resistivity of electron beam welded joint resistivity was higher than that of copper base metal, thus, the electron beam welding method reduces the conductivity of copper. As the gas pore, crack, and other defects reducing in the joints with stirring relatively fierce in the triangle wave scanning mode, the weld structure is a new distorted fine equiaxed grain, the lattice type is the same as the parent material and the composition is the same, so the resistivity is closest to the parent material. To summarize, The main reason for the decrease of conductivity is a degree of grain refinement occurred after electron beam welding, the number of grain boundaries increases and the resistivity increases with the increase of grain boundaries, and the grain boundary has a scattering effect on conduction electrons [21], in the process of electron beam welding, impurities can be mixed into the weld, causes the copper

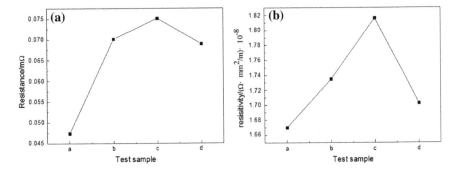

Fig. 8 a Resistance, b resistivity

lattice distortion, when the free electrons flow in the direction, the wave scattering is produced, which leads to the decrease of the conductivity of the weld.

During tensile test, the joint is fractured along the weld zone. The fracture surfaces of the triangular wave scanning mode are analyzed, as shown in Fig. 9. The results of the energy spectrum analysis indicate only Cu element and no other elements or the distribution of the second phase can be detected from the fracture edge (Fig. 9a) and the center of the fracture (Fig. 9b). The fracture morphologies for base metal and joints with non-scanning and triangle wave scanning are shown in Fig. 10. It is shown that there is a layer of the ductile zone in the middle part of the fracture surface, and there are differences in the size of the nest and the tearing edge, which is a typical ductile fracture. An oxide layer can be observed from the fracture, which means the burning phenomenon occurrence, and some gully and some spherical particles near the gully can be observed in Fig. 10c. In Fig. 10d, the fracture surface of the specimen is stratified by the fracture surface layer, and the layer interface is very obvious, by observing the local amplification Fig. 10f, it can be concluded that the cross-section is relatively flat, and the fracture mode of the joint is a brittle fracture. The plastic flow of metal is more difficult in the joint with non-scanning, which leads to the tensile strength decrease and the brittleness of the joint. The obvious ductile

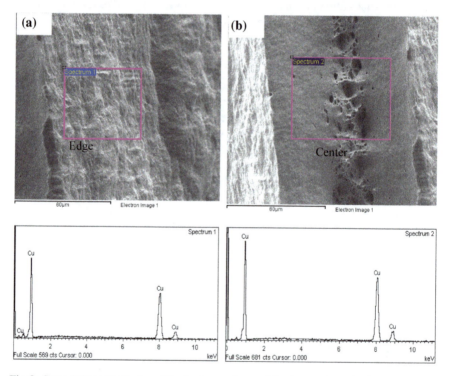

Fig. 9 Second phase distribution of the fracture surface and the sketch map under the triangle wave scanning method

Effect of Scanning Mode on Microstructure ...

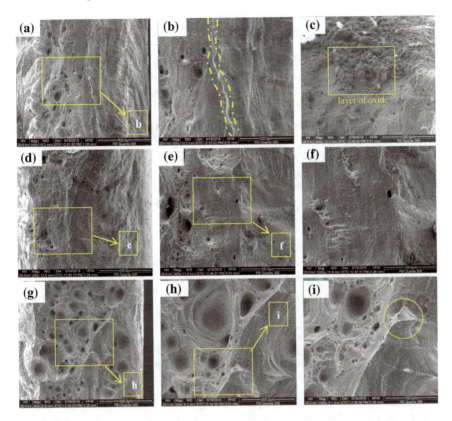

Fig. 10 SEM of tensile fracture morphology: **a–c** base metal, **d–f** non-scanning, **g–i** Triangle wave scanning

fracture characteristics by the local magnification can be found in the fracture surface of joint with the triangle wave scanning mode. The connection mode of aggregation and growth is extended to form a dimple fracture as shown in Fig. 10i. The dimples are characterized by large and deep cave. In Fig. 10e, some tough nest is obviously smaller and shallower than that in Fig. 10h. The size of the dimple mainly depends on the size or quantity of the inclusions and the second phase particles. The depth of the dimple characterizes the plastic of metal. The deeper the dimple is, the better the plastic is, and vice versa. Based on the tensile results, the tensile strength and elongation of the triangular wave scanning joint are higher than that of the joint with non-scanning and other scanning waveforms. It is proved that the triangular wave scanning changes the fracture mechanism of the joint, and the joint is changed from brittle fracture to ductile fracture, which improves the mechanical properties of the joint.

4 Conclusions

(1) The copper joint can be welded with vacuum electron beam with different scanning model, and the vigorous stirring of scanning model for metal pool is benefit to the formation of free defect joint, improving the mechanical properties of the joint.
(2) Due to the difference in the energy density of different scanning modes, the mechanical properties of welded joints with different scanning methods are different. The average tensile strength of joints is up to the 178.8 MPa, near 85% of the tensile strength of base metal.
(3) The mechanical properties of joints different as the different scanning models, i.e., a high fracture toughness for joint with triangular wave scanning, a larger ductility for the joint with triangular wave scanning.

References

1. Huang XH, Tu JP, Zhang CQ et al (2007) Spherical NiO-C composite for anode material of lithium-ion batteries. Electrochim Acta 52(12):4177–4181
2. Poizot P, Laruelle S, Grugeon S et al (2002) Rationalization of the low-potential reactivity of 3d-metal-based inorganic compounds towards Li. J Electrochem Soc 149(9):A1212–A1217
3. Poizot P, Laruelle S, Grugeon S et al (2000) Nano-sized transition-metal oxides as negative-electrode materials for lithium-ion batteries. Nature 407:496–499
4. Neudecker BJ, Zuhr RA, Bates JB (1999) Lithium silicon-tin oxynitride (LiySiTON): high-performance anode in thin-film lithium-ion batteries for microelectronics. J Power Sources 81–82:27–32
5. He DQ, Wu HG (2010) The friction stir welding of copper with a large thickness. J Univ Sci Technol Beijing 10(32):1302–1305
6. Irisarri AM, Barreda JL, Azpiroz X (2010) Influence of the filler metal on the properties of Ti-6Al-4V electron beam weldments: part I: welding procedures and microstructural characterization. Vacuum 84:393–399
7. Zhang M, Chen G, Zhou Y et al (2014) Optimization of deep penetration laser welding of thick stainless steel with a 10 kW fiber laser. Mater Des 53(1):568–576
8. Sokolov M, Salminen A, Katayama S et al (2015) Reduced pressure laser welding of thick section structural steel. J Mater Process Technol 219:278–285
9. Li S, Chen G, Zhou C (2015) Effects of welding parameters on weld geometry during high-power laser welding of the thick plate. Int J Adv Manuf Technol 79(1):177–182
10. Powell J, Ilar T, Frostevarg J et al (2015) Weld root instabilities in fiber laser welding. J Laser Appl 27(S2)
11. Bachmann M, Avilov V, Gumenyuk A et al (2014) Experimental and numerical investigation of an electromagnetic weld pool support system for high power laser beam welding of austenitic stainless steel. J Mater Process Technol 214(3):578–591
12. Wu Y, Cai Y, Wang H et al (2015) Investigation on microstructure and properties of the dissimilar joint between SA553 and SUS304 made by laser welding with filler wire. Mater Des 87:567–578
13. Zhang BG, Wu L, Feng JC (2004) Research status of electron beam welding technology at home and abroad. Welding 2:5
14. Hu HQ, Feng CS (2007) Metal solidification principle. Machinery Industry Press, Beijing

15. Nakagawa H, Katoh M, Matsuda F (1970) Effects of welding heat and travel speed on the impact property and microstructure of FC welds. Trans JWS 1:94–98
16. Huang XY (2008) Electron microscopy analysis of material microstructure. Metallurgical Industry Press, Beijing
17. Xing L, Hang CP, Ke LM et al (2002) Microstructure and electrical properties test of friction stir welded joint of copper. Proc Eleventh Nat Weld Conf 1:456
18. Callister DW Jr (2000) Materials science, and engineering, an introduction. Wiley, New York
19. Lu L, Shen YF, Chen XH et al (2004) Ultrahigh strength and high electrical conductivity in copper. Science 422–426
20. Xue P, Xiao BL, Zhang Q et al (2011) Achieving friction stir welded pure copper joints with the nearly equal strength to the parent metal via additional rapid cooling. Scripta Mater 64(11):1051–1054
21. Zhao XJ (2016) Study on fiber laser welding T2 copper plate and joint performance. Liaoning University of Technology

Ziyang Zhang is Welding Engineer, got the master's degree at Nanchang Hangkong University in 2018.6, and now works at Nanjing Engineering Institute of Aircraft Systems at 2018.7.

Interesting research are the development of special welding technology, the joint of aluminum, superalloys, and metallic glass alloys by electron beam welding (EBW) and laser beam welding (LBW), and stress and strain of weld structure and surface modification with EBW and LBW.

Shanlin Wang is Associate Professor, got the Ph.D. in materials science and metallurgy in Kyungpook National University of Korea in 2011, and then worked at School of Aeronautic Manufacturing Engineering of Nanchang Hangkong University at 2011.3. During 2012–2013, he worked at postdoctoral center of materials science and engineering in Shanghai Jiaotong University. And now, he works in School of Aeronautic Manufacturing Engineering at Nanchang Hangkong University. Now, preside and attend 4 National Natural Science Foundations of China, and 3 Jiangxi Provincial Natural Science Foundation of China; the number of academic papers is over 30.

Interesting research are the development of Fe-based metallic glass and its industrial application, the joint of titanium, aluminium, superalloys, and metallic glass alloys by electron beam welding (EBW) and laser beam welding (LBW), and stress and strain of weld structure and surface modification with EBW and LBW.

Research on Virtual Reality Monitoring Technology of Tele-operation Welding Robot

Canfeng Zhou, Long Wang, Yu Luo, Hui Gao, Juan Li and Guoxue Gao

Abstract In order to replace human to carry out welding tasks in hazardous environment such as underwater, space, nuclear plant, and underground, a tele-operation welding robot system based on virtual reality technology was constructed, which is comprised of slave robot and master virtual working environment. Various aspects of the tele-operation welding robot were discussed, such as robot construction, 3D model building and application in Unity, network communication between the virtual robot and the physical robot, and tele-operation welding test trial. The working environment created by VR technology helps the operator to control the slave robot and conveniently view 3D models of the slave robot and the welding equipment. Welding parameters including welding current, voltage, and arc also can be collected and shown in the master side. The 3D virtual master robot can accurately catch dynamical position of the physical slave robot in real time, and notify operator on the screen when collision between welding gun and workpiece occurs.

Keywords Tele-operation · Welding robot · Virtual reality · Network communication · Unity

1 Introduction

As main content of remote science system, tele-operation supported by multi-technologies is very important in robot research. Tele-robot plays an important role to perform tasks in hazard or inaccessible environment. The first remotely controlled robot in space named ROTEX was developed in 1993 [1]. A unified tele-operated autonomous dual-arm robotic system was built as part of a tele-robotics program [2]. Engineering Test Satellite VII carrying a 2 m long 6-Degree of Freedom (DOF) tele-operated robot arm was launched in 1997 [3]. A tele-robotic system was devel-

C. Zhou · L. Wang · Y. Luo (✉) · H. Gao · J. Li · G. Gao
Beijing Higher Institution Engineering Research Center of Energy Engineering Advanced Joining Technology, Beijing Institute of Petrochemical Technology, Beijing, China
e-mail: luoyu@bipt.edu.cn

© Springer Nature Singapore Pte Ltd. 2019
S. Chen et al. (eds.), *Transactions on Intelligent Welding Manufacturing*,
Transactions on Intelligent Welding Manufacturing,
https://doi.org/10.1007/978-981-13-7418-0_9

oped to perform the complex deactivation and decommissioning activities in nuclear plant [4]. The concept of telepresence is to put forward in supervisory control of telerobotic system, which means that the operator receives sufficient information about the tele-robot and the task environment, displayed in a sufficiently natural way that the operator feels physically present at the remote site [5]. In fact, great progress of VR technology accelerates the realization of telepresence more effectively, accurately, and low costly. Application of VR technology offers the possibility of performing remote operation with greater safety and comfort [6–8]. A tele-operation construction robot control system with VR was developed to perform the remote operation of slave robot by manipulating the graphic robot directly in virtual environment using joysticks [9]. A new 3D virtual environment modeling technology was proposed to realize accurate operation of space tele-robot and to minimize the effect of time delay [10].

2 System Structure of the Tele-operation Welding Robot Based on VR Technology

The tele-operation welding robot is divided into two parts named the master system and the slave system, which is shown as Fig. 1. Here, the slave system is comprised of a 6-DOF welding robot driven by servo motors and controlled by PMAC motion control card through a computer (PC1), a positioner to supply welding position required in task, a welding source and a welding camera. The master system is controlled by an operator through a control box, a 3D working environment created with VR technology and displayed in screen of another computer (PC2). The operator performs remote operation of the welding robot by using the control box linked with PC1 and 3D virtual working environment running on PC2. Motion control information from the control box is input into PC1 and sent to PMAC motion control card, which calculates operational signals for the servo drivers to drive six servo motors of the robot. In the developed master system, the operator can from all directions to view CG (Computer Graphics) images of the remote welding robot and the task object including the positioner, the wire feeder, and the welding gun. CG images are updated dynamically with posture change of slave robot through high-speed data transmission supported by the LAN (Local Area Network) between PC1 and PC2. The operator not only can control the slave robot in real time, but also can simulate a task in the 3D working environment before performing, which is called off-line programming or off-line teaching. During welding process, welding information such as welding current, voltage, and arc can be collected and analyzed supported by arc welding information analyzer, data acquisition card, and some matching software.

Fig. 1 Structure of the tele-operation welding robot based on VR technology

3 Development and Control of the 6-DOF Robot and Integration with Welding Equipment

The slave 6-DOF robot is made up of mechanical body, control cabinet, and control box. Mechanical body has six joints driven by servo motors, which can supply 0.05 mm repeated positioning accuracy and 5000 r/min maximum rotation speed. In order to run the robot safely, a series of measures were taken, such as emergency stop button, joint overload protection, encoder disconnection protection, and limit switch. A PMAC motion control car is used in the robot to receive command from PC1, calculate operational signal for servo driver, and also at the same time sample rotation angle of servo motor from photoelectric encoder is sampled and transferred to PC1. PMAC is a card designed according to DSP (Digital Signal Processor), which has fast computing speed and large memory space to support high-level motion control required by the robot.

3.1 Welding Data Acquisition

In order to carry out welding, the robot is integrated with several welding equipment by I/O or other means. The welding source is an outstanding digital insulated-gate bipolar transistor (IGBT) product, which can supply welding current with perfect

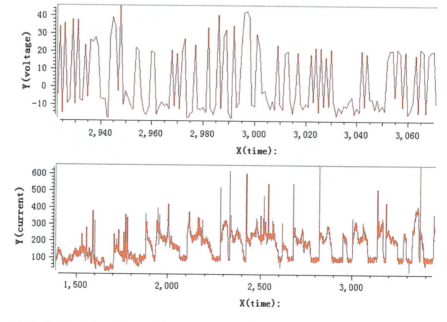

Fig. 2 Welding information acquisition interface

waveforms needed in the gas metal arc welding (GMAW) process. The welding gun is carried by the end joint of the robot to realize welding movement trajectory. Arc welding information analyzer is used to supply normalized welding voltage and current of the welding source for the data acquisition card, which can be connected with PC2 by USB and programmed to control by the calling of Application Programming Interface (API) in the form of static linking library. The welding camera near the welding gun captures images of arc and molten during welding process and transfers image data to PC2 through IEEE 1394 cable. Workpiece is clamped on the positioner, which has the ability to rotate within 360° and turn over within ±90°. Welding data are displayed in an interface as shown as Fig. 2 which is programmed using C++ in a cross-platform development tool.

3.2 Function Design of the VR Monitoring Environment

The VR monitoring environment is made up of four function modules listed in Fig. 3. Communication module is to build connection between the master system and the slave system to transfer information such as control command, robot position, and mission name. Client/Server (C/S) mode and Transmission Control Protocol/Internet Protocol (TCP/IP) are used in network communication. Posture synchronization between the virtual robot and the physical robot is realized, and the

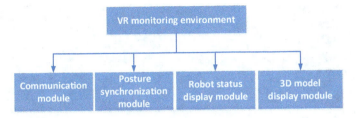

Fig. 3 Function modules of the VR monitoring environment

synchronizing cycle is set as 10 ms in the research but which can be changed according to requirement of application. Robot status display module is in charge of all HMI (Human Machine Interface) functions such as joint rotation speed, control buttons, and camera operation. 3D model display module is to carry out model construction and application of robot and all welding equipment.

3.3 Realizing of the VR Monitoring Environment

3.3.1 Building and Loading of 3D Models of the Robot and Welding Equipment

When coordinate system and measurement unit are determined, 3D models of robot components with same sizes as slave physical robot are built using SolidWorks, which includes robot base, robot waist, tilting big arm, tilting forearm, rotating wrist, and tool joint with pitching axis and rolling axis. These 3D models are imported into 3D Studio Max to improve model precision and adjust model coordinate axis, and to be transformed to a FBX format file which can be loaded in Unity, a cross-platform and all-purpose game engine that supports 2D and 3D graphics, drags and drops functionality and scripting through C#.

Those former FBX files of robot components exported from 3D Studio Max are loaded and managed as a source type named "Prefab" in Unity, which can be created, deleted, and cloned dynamically in program execution. These 3D models of robot components listed in the Unity project catalog are demonstrated in Fig. 4, which can be displayed in a preview mode shown in Fig. 5 and can be programmed to perform a series of operations such as displacement, rotation, scale, and color. Four types of light are supported by the Unity to bring better telepresence for objects in the scene, and parameters of each type can be set conveniently, such as size, color, intensity, and shadow.

Fig. 4 3D models of robot components listed in the Unity project catalog

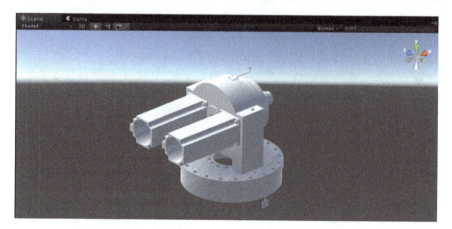

Fig. 5 Model files in preview mode

3.3.2 Network Communication Between the Virtual Robot and the Physical Robot

Reliable and high-speed data transfer through LAN between the virtual robot and the physical robot, which is required to realize CG image update, off-line teaching and real-time control. Network communication between the virtual robot and the physical robot is established by a series of technologies, such as computer network based on C/S, communication protocol of TCP/IP, C++ language, and Socket interface, which is a Winsock API applied in Windows system to accomplish network programming.

The TCP/IP protocol uses a four-layer reference model. The transportation layer of the model can be realized by TCP or UDP (User Datagram Protocol). Although network consumption of TCP is higher than UPD, but it can implement correct data transmission independently and do not need additional verification method defined by the user in the application layer. In order to realize motion synchronization between the virtual robot and the physical robot accurately and dynamically, and to avoid inevitable interference in the site, TCP/IP is selected as communication protocol.

In the C/S mode, the side sending the command is called client, and the side responding to the command is called server. As structure of the tele-operation welding robot is shown in Fig. 1, the server keeps monitoring the LAN Interface of PC1 during the running of the physical robot, establishes connection between PC1 and PC2 when request from the client of PC2 is received, and sends motion information of the physical robot to the client such as moving speed and position of each joint.

The communication between the virtual robot client and the physical robot server is realized by transmission of character array which represents a piece of information. Communication information of the tele-operation welding robot can be divided into three types related to the physical robot, which are robot running status, robot motion, and robot controlling command. Robot running status includes a series of information, such as the status of PMAC card, the robot running mode, the mission name of the robot to perform, and the error information sent by the robot.

Programming of the network communication between the virtual robot client and the physical robot server is realized by Socket interface, which is a Winsock API applied in Windows system to accomplish network programming. Winsock API contains some library functions, which can be used in the TCP/IP protocol to receive and send network data. These library functions are called by applications developed on Windows platforms to perform operations. Process of network communication using Socket is shown in Fig. 6. The server firstly establishes Socket and binds with the IP and the port, secondly, begins to monitor the port to receive data. When the Accept ( ) function is called, the thread in charge of the flow running will go into a blocked status, and the communication between the virtual robot client and the physical robot server will be started until there is a client connecting to the server.

3.3.3 Programming of the VR Monitoring Environment Based on Unity

Three languages running on the .Net platform based on the Mono are supported by Unity, which are C#, JavaScript, and Boo. The overview program flow chart of the tele-operation welding robot is shown in Fig. 7. Program initialization is executed to load the scene, the robot model, the surrounding objects and the light. If the physical robot data is read successfully, the virtual robot and HMI will be updated, otherwise error message will be printed. Condition of the physical robot is always monitored during all the program running time to detect robot running status abnormal such as collision, which is needed to send stopping command to robot. Besides that, program will also detect the action of the operator through the HMI and send command to the robot according to the operation.

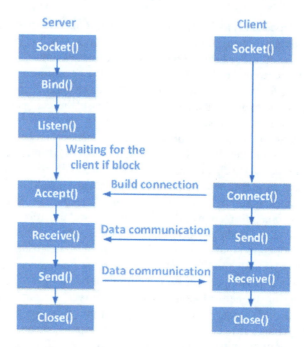

Fig. 6 Process of network communication using Socket

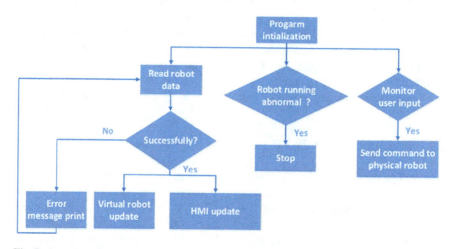

Fig. 7 Overview program flow chart of the tele-operation welding robot

3.4 Posture Synchronization Between the Virtual Robot and the Physical Robot

Posture synchronization is realized by update the virtual robot frame by frame according to data gathered from the physical robot. A library function named FixedUpdate ( ) is called to perform updating operation, which is a function constantly executed in a fixed time interval during the entire Unity life circle and will not be affected by the computing capacity. All related methods of updating the virtual robot are stored in the FixedUpdate ( ) function, and the running interval can be modified.

In the VR monitoring environment, all joints of the virtual robot will be loaded in the scene. Position and posture of an object in a three-dimensional space are described by six variables, they are the coordinate of the x- , y- , z-axis, and the rotation angle around the corresponding axis. Because each joint of the tele-operation welding robot rotates around a fixed axis, so synchronization between the virtual robot and the physical robot is equivalent to rotation angle synchronization between the virtual axis and the physical axis.

3.5 Detection of Robot Collision

There are two operations to carry out collision detection of the tele-operation welding robot in the VR monitoring environment, one is to change the robot color of the corresponding model when the distance of the robot to the surrounding objects is less than a set value, another is to stop the robot and change the robot color of the corresponding model when the robot collides to the surrounding objects. The hierarchical bounding box algorithm is used in Unity to perform collision detection, which will call one of three functions named OnTriggerEnter ( ), OnTriggerStay ( ), and OnTriggerExit ( ) according to different related position status between analyzed bounding boxes.

4 Tele-operation Robotic Welding Test

4.1 Welding Test System and Welding Process

The real tele-operation robotic welding test system including all components is shown in Fig. 8. Using the control box, the operator can accomplish single axis moving in the joint coordinate system or rectilinear moving in the rectangular coordinate system. Testing program is prepared in teaching mode and carried out in repeating mode. Virtual cameras set in the VR monitoring environment can help the operator to have a better observation of the physical robot, such as position and posture, relationship with surrounding objects, and collision reminding.

Fig. 8 Tele-operation robotic welding test system

Fig. 9 Welds in horizontal position and vertical position

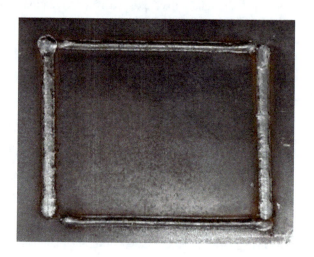

During the welding test, the positioner is set in the vertical position to satisfy requirement of horizontal welding and vertical welding. Not only welding current and voltage of the welding source, FastMig KMS 450, are sampled and shown using waveforms in PC2, but also welding arc and molten pool are captured and displayed. GMAW process is executed in test to weld the workpiece material of Q345B using a diameter of 1.2 mm solid wire classified as AWS A5.18 ER70S-6, and mixed gases as protection with ingredients of 80%Ar + 20%CO_2. A series of tele-operation robotic welding tests are carried out, which indicates that the physical robot runs reliably and the welding process is maintained stably. Welds are shown in Fig. 9 are produced as welding current of 169 A, welding voltage of 25.5 V, welding velocity of 40 mm/s, and gas flow rate of 20 L/min.

Fig. 10 Posture comparison between the virtual robot and the physical robot in one position

Fig. 11 Collision function of the welding gun

4.2 Verification of the Synchronization Function

Verification of the synchronization function is realized by posture comparison between the virtual robot and the physical robot during the running process. Comparison result is shown in Fig. 10, which indicate that they have exactly same postures. Movement of the physical robot can be accurately and dynamically recreated in the VR monitoring environment, because rotation angles of all joints of the physical robot are captured and transferred to the PC2 to update 3D images of the virtual robot in real time. Because all 3D models in the VR monitoring environment have same shapes and sizes as physical objects, and the virtual robot can be viewed from different angles and distances, so the operator can finish the robot teaching process off-line through the HMI integrated with the camera graph.

4.3 Verification of the Collision Detection Function

As shown in Fig. 11, when the welding gun is very close to the workpiece during the teaching operation, the collision detection program finds the phenomenon and changes the welding gun color to yellow to remind the operator.

5 Conclusion

In this paper, a tele-operation welding robot based on VR technology was developed and a tele-operation mode welding test was carried out. Several conclusions are drawn as follows:

- Because of a large sum of money to support life safety for welder can be saved, the tele-operation welding robot system can reduce the cost of repair welding in hazardous environment significantly, and the flexibility of the robot system is effectively improved by the VR technology.
- The 3D virtual robot can catch dynamical position of the physical robot in real time and accurately and give remind to operator on the screen when collision between welding gun and workpiece happens.
- The 3D virtual robot can be viewed from different angles and distances, which can give more comprehensive and detailed information to operator.
- The physical robot can be controlled through real-time mode, or off-line mode supported by 3D VR software, and welding information such as current, voltage and arc image can be collected and analyzed in the tele-operation welding robot system.

Acknowledgements This work is supported by the Natural Science Foundation of Beijing (KZ201210017017).

References

1. Hirzinger G et al (1994) Sensor based space robotics-ROTEX the first remotely controlled robot in space. In: Proceedings of the 1994 IEEE international conference on robotics and automation, San Diego, pp 2604–2611
2. Hayati S, Lee T, Tso K et al (1991) A unified teleoperated autonomous dual-arm robotic system. IEEE Control Syst 11(2):3–8
3. Oda M (1997) System engineering approach in designing the teleoperation system of the ETSVII robot experiment satellite. In: Proceedings of the 1997 IEEE international conference on robotics and automation albuquerque, vol 4, New Mexico, pp 3054–3061
4. Noakes MW et al (2002) Telerobotic planning and control for DOE D&D operations. In: Proceedings of the 2002 IEEE international conference on robotics and automation, Washington, pp 3485–3492
5. Stassen HG, Smets GJ (1997) Telemanipulation and telepresence. Control Eng Pract 5(3):363–374
6. Yamada H, Kato H, Muto T (2003) Master-slave control for construction robot teleoperation. J Robot Mechatron 15(1):54–60
7. Zhao D, Xia Y, Yamada H et al (2003) Control method for realistic motions in a construction tele-robotic system with a 3-DOF parallel mechanism. J Robot Mechatron 15(4):361–368
8. Yamada H, Muto T (2003) Development of a hydraulic tele-operated construction robot using virtual reality—new master-slave control method and an evaluation of a visual feedback system. Int J Fluid Power 4(2):35–42
9. Tang X, Yamada H (2011) Tele-operation construction robot control system with virtual reality technology. Proc Eng 15:1071–1076
10. Zhu BY, Song AG, Xu XN (2015) Research on 3D virtual environment modeling technology for space tele-robot. Proc Eng 99:1171–1178

Canfeng Zhou was born in 1970, received Doctor of Technical Science at 1998, and was promoted to professor at Beijing Institute of Petrochemical Technology at 2008. Mr. Zhou has been studying underwater welding and welding automation, and now is the vice director of Beijing Higher Institution Engineering Research Center of Energy Engineering Advanced Joining Technology. As a main research member, Mr. Zhou has participated in several big projects including projects supported by The National Natural Science Foundation of China (NSFC), the National High Technology Research and Development Program ("863"Program). Research information of Mr. Zhou can be found from his more than 100 published papers.

Mr. Yu Luo was born in 1981, received Doctor of Technical Science at 2012, and was promoted to associate professor at Beijing Institute of Petrochemical Technology at 2016. Mr. Luo has been studying welding automation and joining technology, and now is the key members of Beijing Higher Institution Engineering Research Center of Energy Engineering Advanced Joining Technology as a main research member, Mr. Luo has participated in several big projects including projects supported by the National Natural Science Foundation of China (NSFC), the National Type footer information here Type header information here High Technology Research and Development Program (863 Program). Research information of Mr. Luo can be found from his more than 20 published papers.

The Influence of Powder Layers Intervention on the Microstructure and Property in Brazing Joints of Titanium/Steel

Pengxian Zhang, Yibin Pang and Shilong Li

Abstract Brittle intermetallic compounds in brazing joint of titanium alloy and stainless steel easily make the weld organization and mechanical properties worsen. In order to solve the above problem, tests were performed to compare the different types of metal powder layers in the brazing joints of titanium/steel. On this basis, the mechanical property and microstructure of these joints were explored as technology parameters were changed and different powders were added. The test results show that the elements such as Nb, Cr, V in the powder layer participate in the metallurgical reaction of joints so that the newly born phase appears in joints. At the same time, the microstructure of joints was transformed into the mixture of metal powder particles, solid solution, solder, and various brittle phases. This makes mechanical properties of joints clearly better than the joint of non-powder layer. At the joints where Cr powder layer was added beside the stainless steel and Nb–V powder layer, was added beside the titanium alloy simultaneously, and their maximum shear strength reached 193 MPa, which is more than twice of the joints without metal powder layer. It is the main reason of the improving mechanical property that newly born phase of Cr–Fe, Ti–V replace pervious Fe–Ti and Cu–Ti IMCs in microstructure of the joints.

Keywords Brazing of titanium/steel · Metal powder · Solid solution · Microscopic structure · Mechanical properties

1 Introduction

The combined components consisting of titanium alloy and stainless steel can combine the advantages of the two materials and also have good economic benefits, which have been found good applications in the important fields of aerospace and national economy [1]. However, the welding of titanium alloy and stainless steel

P. Zhang (✉) · Y. Pang · S. Li
Gansu Provincial State Key Laboratory of Advanced Processing and Recycling of Non-ferrous Metals, Lanzhou University of Technology, Lanzhou 730050, China
e-mail: pengxzhang@163.com; pangyibin1995@163.com

generates a large amount of titanium/iron brittle intermetallic compounds due to the significant mismatch in the physical and mechanical properties of two metal materials, resulting in deterioration of the joint performance of the components, which is the basic problem that the great majority dissimilar metals need to be solved for reliable connection [2]. In order to solve the issues mentioned above, it is an effective way to add a transition metal layer during solid phase welding for achieving a reliable connection of titanium/steel. Zhao et al. [3] placed a 1 mm thick germanium plate and copper plate between a titanium alloy and stainless steel for vacuum hot-rolling welding. As a result, the interlayer metal restrained the formation of brittle phases and the joint performance was significantly improved. Lee et al. [4] added a metal foil layer of Cr/V/Ni composites between industrial pure titanium and stainless steel, having simultaneously added Ti-based amorphous solder to heat the joint in an infrared manner to implement their reliable connection. Li et al. [5–7] studied the titanium/steel connection when the metal foil, electroplated layer, and sprayed layer were used as intermediate transition layers. Researches have shown that transition metal elements are effective in improving the microstructure and properties of joints. Furthermore, owing to the addition of solid metals, the interface that is artificial to increase is the weak link of the joint strength. However, it is of great key to reliable connection of titanium/steel about which intervention ways of transition metal elements are chosen to use. Additionally, the use of multi-alloy brazing alloy as a transition layer can significantly reduce the influence of the interface on the mechanical properties, but the components of brazing alloy that can be used in the industry are still relatively single and cannot meet the requirements for the reliable connection of titanium/steel, the composition and metallurgical behavior of which need further study [8, 9].

Based on the above review of literatures, this paper attempts to use the resistance brazing of titanium/steel of the alloy powder layer and the brazing material as the intermediate transition layer, studying the formation mechanism of the brazed joint of titanium/steel after the intervention of different metal powder layer.

2 Experimental Details

Commercially available TC4 (Ti–6Al–4V) titanium alloy and 304 stainless steel (1Cr18Ni9Ti) were used as experimental materials, which were sliced into 80 mm × 20 mm × 1.5 mm rectangular solids by wire-cutting for brazing. The brazing filler metal was BAg45CuZn silver-based brazing filler metal with a thickness of 0.2 mm. The metal powder was selected from Cr, Ni, V, and Nb powders with a particle size of 400 mesh, of which the purity reaches up to 99.9%. The structural sealant of YK-8907 1300 °C single-component ultra-high temperature was used for the oxidation-proof sealant. This test used lap joints. Before the welding, the faying surfaces of the base material were first polished with sandpaper up to grit 800 to remove the oxide layer, ultrasonically cleaned in acetone for more than 3 min to remove residual impurities and oil stains and then assembled as the joint.

The Influence of Powder Layers Intervention …

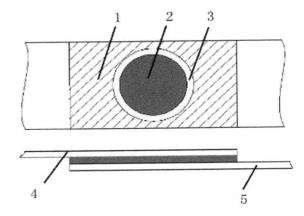

Fig. 1 Diagram of joint assembly. (1) Sealant, (2) metal powder and solder, (3) reserved space, (4) titanium alloy, (5) stainless steel

The diagram of the joint assembly is shown in Fig. 1. At first, the solder sheet was punched into a wafer with 14 mm in diameter, grounded by a 1000-grit sandpaper step-by-step to remove the oxide layer, and then ultrasonically cleaned. Then, the metal powder was evenly spread into a round groove in diameter of 15 mm on the surface of the special mold. After the metal powder was placed in the solder sheet, the manual press was used to press the powder on the surface of the solder. At the same time, the brazing sheet with metal powder pressed was placed on the specific position of the stainless steel surfaces which shown in the second position in the figure. Next, when the high-temperature structural adhesive was evenly applied on the first position of the figure, a certain clearance should be reserved around the solder, for which the structural adhesive is prevented from contaminating the brazing area during assembly and welding. The certain clearance was shown in the third position in the figure. Finally, the titanium alloy was assembled according to the form shown in Fig. 1, and the second position of the parent material was pre-pressed with a press to ensure that the parent material, the solder and powder were reliably contacted and reliably energized during welding.

Table 1 Combination mode of brazing joint

The types of brazing joint	The combination mode of brazing joint	The mass ratio of mixed powder
Type I	SS + B + TC4	–
Type II	SS + B + (Nb, V) + TC4	Nb:V = 17:33
Type III	SS + Cr + B + (Nb, V) + TC4	Nb:V = 17:33
Type IV	SS + (Cr, Ni) + B + Nb + TC4	Cr:Ni = 1:1
Type V	SS + (Cr, Ni) + B + (Nb, V) + TC4	Cr:Ni = 3:1, Nb:V = 3:1

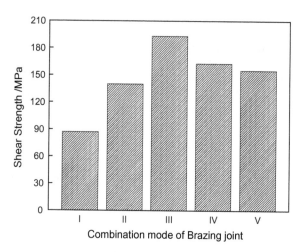

Fig. 2 Tensile results of different brazing joint

The experiment selected a method for simultaneously adding different kinds of metal powders with 0.02 g in weight to the interface between the solder (B) and the titanium alloy (TC4), and between the solder and the stainless steel (SS). Five kinds of titanium/steel brazed joints I, II, III, IV, and V were designed experimentally, for which mechanical tests and microstructure analysis were performed after welding. Five types of joint combinations and the ratio of metal powders are listed in Table 1.

3 Effect of Powder Layer Interference on Mechanical Properties of Joints

In order to evaluate the joint performance under resistance brazing, tensile tests were used as a means of evaluating the mechanical properties. The resistance brazing process tests were performed on assembled titanium/steel joints with reasonable welding current, welding time, and electrode pressure parameters explored. The resistance brazing of different types of titanium/steel joints was performed under reasonable parameters, and then five types of joints were tested and analyzed. Figure 2 shows the tensile results of the above five types of joints. The measured value of each test is the average value of the three test results under the same conditions. According to the figures, it can be concluded that the joint strength with the metal powder added is significantly greater than that of the joint without the metal powder under the same experimental conditions. Among the joints added with metal powders, the maximum shear strength of brazed joints was obtained as 193 MPa corresponding to the Type III joint, and the Type II joints were the worst with only 140 MPa. Shear strength of the type I joint without powder intervention only is 87 MPa, which is much lower than that of the joint with added metal powder. Therefore, the addition of metal powder can effectively improve the performance of joints in titanium/steel brazed joints.

4 Effect of Powder Layer Interference on Microstructure of Joints

In the brazing process, the brazing filler material melts and flows at the joint, ultimately forming the brazing joint by cooling and solidification. Changing the wettability and spreadability of the surface of the base metal, the intervention of metal powders changes simultaneously the metallurgical reaction at the joint. Through the observation of the microstructure of the joint and the analysis of the fracture, the influence mechanism of the intervention of the metal powder on the formation of the titanium/steel brazed joint can be explored. Figure 3 shows the microstructure of different types of joints. It can be seen that the materials near the interface are divided into three different zones, that is, the left side is TC4 titanium alloy; the right side is 304 stainless steel; and the middle is the brazing seam zone.

XRD was used to further confirm the reaction components for different types of joints and the results are given in Fig. 4. The location where the XRD measurements were taken is in cross section of the joints. From in Fig. 4a, when the metal powder was not added, the solder joint area was obvious, and a large number of gray block-like structures and some dendrite structures can be observed near the titanium alloy side. The point scanning showed that the main elements are Cu and Ti, which can be determined as Cu–Ti intermetallic compounds. It is revealed that in the process of brazing, the Ti element diffused to the opposite side and formed brittle phase with the Cu element in the joint, and the reaction layer was the weak zone of the joint. In the type II joint structure of Fig. 4b, since only a mixed powder of Nb and V was added on the side of the titanium alloy, a large number of gray clusters of Ti, Nb, V solid solution and Cu–Ti, Nb–Zn system brittle phase, and a small amount of metal powders that had not been reacted, appeared in the brazing zone near the titanium alloy side. In the solder region near the stainless steel side, there was a small amount of gray block-like structure and dendrite, which is a Cu–Ti brittle phase of mainly composed Ti_2Cu formed by the reaction between the unblocked Ti and copper in the solder. With white solder material in the solder joint area of the type III joint largely disappeared, a large amount of gray cluster-like structures appeared near the titanium alloy side and the stainless steel side were unreacted Cr powder particles as shown in Fig. 4c. At the same time, the Cu–Ti brittle phase in the joint was further reduced, and more mixture of Ti, Nb, V solid solutions and NbZn, $NbZn_2$ compounds were appeared in the solder zone and FeCr compound appeared on the stainless steel side. Moreover, the white solder in the solder joint area of the Type IV joint was further decreased with a large amount of gray lump structures and a small amount of unreacted metal powders appeared, as seen in Fig. 4d. When Nb_6Zn_7 compound and Nb, Ti solid solution were formed close to the titanium alloy side, some substances were formed close to the stainless steel side such as Cr_2Ni_3 compound and Fe, Ni solid solution. On the other hand, Cu–Ti-based compounds and compounds formed from Nb and base metal elements existed in the solder region. In Fig. 4e, the delamination of the solder joint area of the Type V joint was not obvious, and there was a small amount of unreacted Nb and Cr powder particles in the joint. In the joint structure,

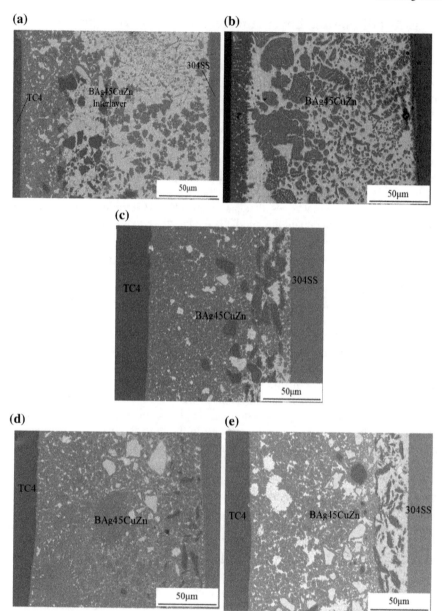

Fig. 3 Microscopic structure of brazing joint with different combination mode. **a** Type I joint, **b** type II joint, **c** type III joint, **d** type IV joint, **e** type V joint

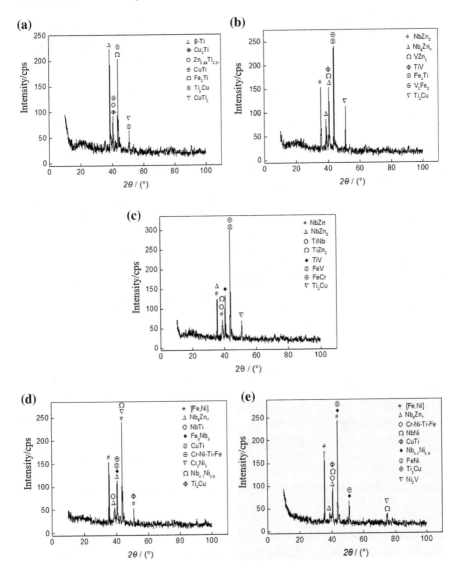

Fig. 4 X-ray diffraction analysis of different brazing joint. **a** Type I joint, **b** type II joint, **c** type III joint, **d** type IV joint, **e** type V joint

intermetallic compounds NbNi, Nb$_{0.1}$Ni$_{0.9}$, and Nb$_6$Zn$_7$ appeared, and Fe, Ni solid solution and compound FeNi were generated near the stainless steel side solder area.

It can be seen from the X-ray diffraction results that the type and amount of the substances in the titanium/steel brazed joints are significantly changed when the metal powder is added. In addition to the reduction of some harmful intermetallic compounds, a variety of solid solution mixtures occur at the joints, replacing a part of the solder as the joint structure. At the same time, some of the Nb–Zn, V–Zn, and Cr–Fe compounds that are formed exist in the newly formed solid solution. Under these circumstances, the joint strength increased dramatically.

5 Effect of Powder Layer on Joint Fracture

The corresponding fracture surface morphologies on different types of joints are displayed in Fig. 5. It is found that Types I and II joints were fractured in the reaction layer between stainless steel and solder, and the joint of Types III, IV, and V broke in the reaction zone between titanium alloy and powder and solder. The fracture of Type I joint shows a rivers shape pattern with a small quantity of plastic fracture characteristics in the local surface, which is a mixed fracture morphology based on cleavage fracture, indicating that its fracture is brittle fracture. The fracture of the Type II joint exhibits a large number of small and shallow dimples and the tiny torn grain with a uniform distribution, which is ductile fracture. Type III joint fracture presents a mixed morphology of a large number of small rocky-like structures and dimples with torn edges, which has mixed characteristics of ductile fracture and intergranular fracture, suggesting that the joint owns certain plasticity. However, the fracture surface of Type IV or V joint displays short and curved river-like pattern and torn edges, which is quasi-cleavage fracture.

Based on the above analysis, it can be concluded that the wetting and spreading of the brazing material to the base metal have a great influence on the performance of the brazing joint which is also influenced by the strength of the brazing material. After the metal powder is intervened, it reacts with the base metal to strengthen the metallurgical reaction of the joint and generate a variety of new substances. In addition, it is found that the intervention of metal powder can change the distribution of intermetallic compounds. Meanwhile, the brittle phase no longer appears as a patch in the joint but distributes in a small and dispersed manner in the joint structure, which plays a role in strengthening the second phase and enhancing the strength of the solder. At the same time, the intervention of the powder also improves the wetting and spreading of the brazing material to the base metal, so that the connection strength increased. As can be seen from the fracture diagram of the joint, fractures of joints without added powder exhibit typical brittle fractures. After adding the metal powder, the fractured form of the joint becomes a mixed fracture which makes the plasticity of the joint increase. Moreover, the adding methods of different metal powders also change microstructures of the different joints. The fractures of Type III joints are dominated by plastic fractures, of which microstructures are based on brazing filler material with solid solution fine crystals in the core, and it is also clear that the grain size is less than 1 μm in the intergranular fracture region, so the joint strength is greatly improved. Because the fracture of Types IV and V joints show a quasi-cleavage fracture with a large number of brittle phases and a small amount of grain in size of about 3 μm the improvement of the joint strength is not very obvious. To sum up, brazing of stainless steel and titanium alloy, the type of metal powder, the

The Influence of Powder Layers Intervention ... 155

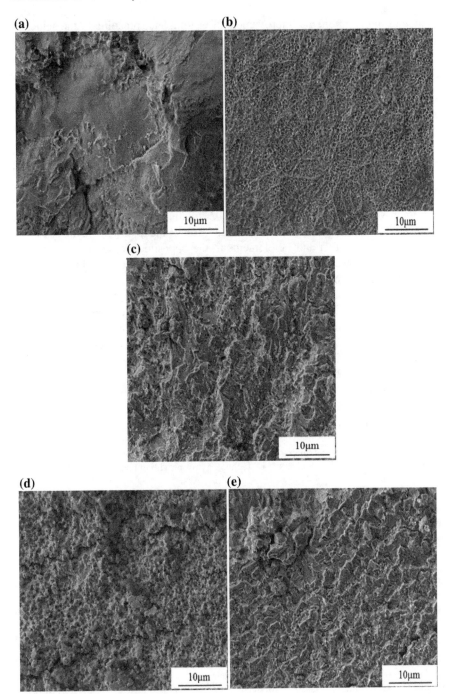

Fig. 5 Fracture morphology of different brazing joint. **a** Type I joint, **b** type II joint, **c** type III joint, **d** type IV joint, **e** type V joint

method of intervention, the improvement of wetting and spreading of the solder by the metal powder and the reaction between the metal powder and the matrix element have a great effect on the performance of the joint.

6 Conclusion

(1) When titanium alloy and stainless steel are brazed, the addition of metal powder layer can effectively improve the mechanical property of the joint. As Cr powder layer was added beside the stainless steel and the mixed powder of Nb and V was added beside the titanium alloy, Type III joint achieved the most apparent improvement in the performance of the titanium/steel brazed joint. It is due to the fact that on the one hand, the addition of metal powder layer can block the diffusion of Ti and Fe elements to the opposite side. On the other hand, the metal powder layer is involved in metallurgical reaction, increasing the metallurgical reaction degree on the interface of the joints, and then V–Ti, Nb–Ti solid solution and Cr–Fe, Nb–Zn compounds replace pervious Fe–Ti and Cu–Ti IMCs in microstructure of the joints. Moreover, the joints exhibit a superior mechanical property compared to the joint of non-powder layer.

(2) The intervention of the metal powder layer can dramatically improve the performance of the joint, and can largely affect the homogeneity of the joint structure. The metal particles cannot be completely covered by the molten solder owing to the influence of its surface energy, resulting in slight declining in the density and easily forming powder clusters and holes in the brazing area which can limit the further improvement in the strength of the joint.

Acknowledgements Special thanks give the National Natural Science Foundation of China (51061011) and the Lanzhou University of Technology for providing financial support.

References

1. Prasanthi TN, Sudha C, Ravikirana et al (2015) Friction welding of mild steel and titanium: optimization of process parameters and evolution of interface microstructure. Mater Des 88(1):58–68
2. Lee MK, Park JJ, Lee JG (2013) Phase-dependent corrosion of titanium-to-stainless steel joints brazed by Ag–Cu eutectic alloy filler and Ag interlayer. J Nucl Mater 439(1–3):168–173
3. Zhao DS, Yan JC, Wang Y et al (2006) Vacuum hot-weld welding of titanium alloy and stainless-steel using copper and tantalum composite interlayer. China J Weld 27(11):99–102
4. Lee MK, Lee JG, Choi YH, Kim DW (2010) Interlayer engineering for dissimilar bonding of titanium to stainless steel. Mater Lett 64(9):1105–1108
5. Li P, Li JL, Xiong JT (2011) Study on diffusion welding process of titanium alloy and stainless-steel added Ni+Nb intermediate layer. J Aer Mater 31(3):46–51
6. Zhang PX, Ma CY, Yu HY (2016) Influencing mechanism of Ni clad layer on atomic diffusion behavior of titanium/steel brazing joints. Rare Metal Mater Eng 45(2):449–453
7. Song XR, Li HJ, Zeng X (2015) Brazing of C/C composites to Ti6Al4V using multiwall carbon nanotubes reinforced TiCuZrNi brazing alloy. J Alloys Compd 664(1):175–180

8. Dong HG, Yang ZL, Wang ZR et al (2014) Vacuum brazing TC4 titanium alloy to 304 stainless steel with Cu-Ti-Ni-Zr-V amorphous alloy foil. J Mater Eng Perform 23(10):3770–3777
9. Elrefaey A, Tillmann W (2007) Microstructure and mechanical properties of brazed titanium/steel joints. J Mater Sci 42(23):9553–9558

Pengxian Zhang works at School of Materials Science and Engineering, Lanzhou University of Technology, Lanzhou 730050, China. Now, he is mainly engaged in new technology and new technology of welding, welding process automation, welding quality control and other aspects of research.

Short Papers and Technical Notes

Process Research on Diode Laser-TIG Hybrid Overlaying Welding Process

Ming Zhu, Buyun Yan, Xubin Li, Yu Shi and Ding Fan

Abstract In order to improve forming quality and reduce equipment cost, a laser-TIG overlaying welding method is proposed. An experiment system was set up aiming to study the welding parameters have influence on forming characteristics. And effects of laser power or TIG current on forming shapes, such as depth, width, dilution rate, and wetting angle have been analyzed. The results show that TIG torch results in the decreased temperature gradient at the edge of welding pool and the improved forming quality; and the increase of TIG current leads to the increase of depth, width, and dilution rate of cladding layer and decrease of aspect ratio and wetting angle.

Keywords Laser-TIG hybrid · Overlaying welding · Forming

1 Introduction

Nickel base alloy powder has been used to prolong the service life and improve production efficiency of equipment by cladding it on the part easy to wear or corrode since it has excellent wear resistance, corrosion resistance, and heat resistance [1]. The study on relation between process parameters and forming characteristics will not only help control overlaying layer quality, but also guarantees the forming dimension precision [2]. Diode laser can precisely control the heat conduction, and compared with other laser, it has large rectangle laser spot which will help obtain cladding layer with low dilution rate and high aspect ratio [3].

However, higher laser power causes several drawbacks including more intensive thermal impact, higher crack tendency, residual stress, and production costs. [4]. Laser coupled with TIG arc can flexibly alter hybrid heat source characteristics so as to meet different heat demands and obtain ideal heat distribution [5]. Most of the

M. Zhu (✉) · B. Yan · X. Li · Y. Shi · D. Fan
State Key Laboratory of Advance Processing and Recycling Non-ferrous Metals,
Lanzhou University of Technology, Lanzhou 730050, China
e-mail: zhumings@yeah.net; 524641909@qq.com

© Springer Nature Singapore Pte Ltd. 2019
S. Chen et al. (eds.), *Transactions on Intelligent Welding Manufacturing*,
Transactions on Intelligent Welding Manufacturing,
https://doi.org/10.1007/978-981-13-7418-0_11

laser-TIG hybrid heat sources are used in deep penetration welding area. Cladding process and cladding layer are improved by applying laser-TIG hybrid heat sources. Thus the method mentioned above can be proved to be promising.

In this paper, in order to prove the effectiveness of TIG arc to laser overlaying welding, the influence of welding parameters on forming characteristics were studied, such as laser power or TIG current to dilution rate, aspect ratio, wetting angle, and width of cladding layer. Process experiments were also carried out to obtain the optimal welding parameters, also a desired welding appearance was acquired.

2 Experimental

Ni60 alloy powder and Q235 plates were chosen as welding material and substrate, respectively, and Ni60 powder was pre-placed on the Q235 plates by a laying powder device.

A 2 kW focus light diode laser (FL-DLIGHT-2000) was used in this experiment. The device parameters include 4 mm × 4 mm laser spot dimension, 976 nm wavelength, and a trapezoid temperature gradient. And the TIG arc to be coupled with laser was ignited by a TIG welding machine. The tungsten electrode with diameter of 3.2 mm used in this experiment was set to be 3 mm away from the work piece, and protecting gas flow was 4 L/min. The TIG arc was followed by laser spot with the distance of 2 mm in cladding process and the angle between the axis of the laser and the arc heat source was 30°. The experimental setup of this system is shown in Fig. 1.

The cross-sectional morphology of cladding layer was obtained by adopting 1.4 kW laser + 50 A TIG arc as shown in Fig. 2. After the specimens were polished and etched by nitric acid alcohol solution, the geometric size of cladding layer

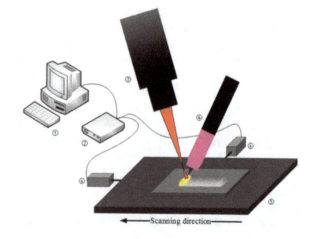

Fig. 1 Schematic of the experimental setup. ① Computer for controlling the work table movement, ② stepper motor driver, ③ diode laser, ④ TIG torch, ⑤ worktable, and ⑥ stepper motor

Fig. 2 Picture of cross-sectional cladding layer

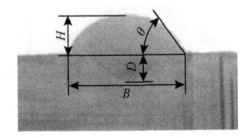

(height of cladding layer H, depth of cladding layer D, width of cladding layer B, and wetting angle θ) can be measured. And then, aspect ratio $a = H/B$, indicating lateral wet ability, and dilution rate $d = D/(H + D)$, indicating longitudinal expansion ability dilution degree of substrate to cladding layer, can be calculated [6]. Wetting angle θ can be calculated as $\sin\theta = (H/B)[(H/B)2 + 0.25]$. In a certain range, θ decreases with the decrease of aspect ratio H/B. The wetting angle and aspect ratio not only reflect the wet ability of the cladding material on substrate, but also reflect whether the cladding process parameters are suitable or not. It is generally considered that $30° < \theta < 60°$ is an ideal wetting angle range [7].

3 Results and Discussion

3.1 The Influence of Laser Power on Laser-TIG Hybrid Cladding Layer Formation

Figure 3 shows the morphology and cross section of the cladding layer obtained by 50 A TIG arc coupled with 0.6, 1.0, 1.4, and 1.8 kW laser, respectively, when the powder thickness was 1.5 mm and scanning speed was 1.5 mm/s. It can be seen that when 0.6 kW laser is used alone, the energy provided by the laser cannot complete the "melt-spheronizer-spread-shrink" process, so the cladding layer is not continuous; when using a 600 W laser coupled with a 50 A TIG arc, the powder can be well clad, indicating that the introduction of TIG arc can effectively reduce the required laser power, and with the increase of laser power, the welding appearance is better.

Figures 4 and 5 show the changing trend of the forming characteristics of single-track cladding layer with laser power under the condition of 50 A TIG arc. Figure 4 indicates that when using laser-TIG hybrid heat source for cladding, the width of the cladding layer increases with increasing laser power; the wetting angle initially decreases with increasing laser power and then basically keeps unchanged. As the laser power increases, the energy absorbed by the molten pool increases, and more substrate material was melted, so that the molten pool area increases and the amount of melting powder increases. Therefore, the width of hybrid cladding layer increases with increasing laser power. The energy absorbed by substrate increases so that the

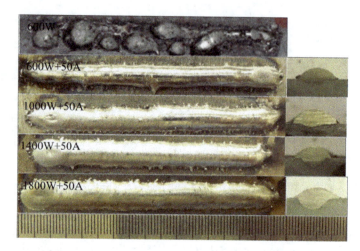

Fig. 3 Appearance and cross-sectional shapes of single cladding with different laser powers in hybrid heat source cladding

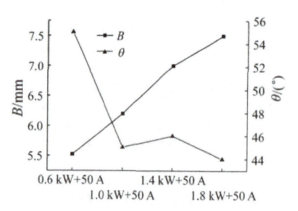

Fig. 4 Effect of different laser powers in hybrid heat source cladding on B and θ

melting amount increases, and then the volume of the molten pool and the penetration increases. The height of cladding layer is affected by the laser power and the powder thickness, and the thickness of the powder plays a decisive role. The height keeps basically unchanged when the powder thickness is fixed. Aspect ratio H/B decreases with the increase of the laser power, and θ decreases with the decrease of the aspect ratio; as the laser power increases, the energy absorbed by the edge of the cladding layer also increases, and the temperature gradient at the edge of the cladding layer can decrease to a further level.

Figure 5 shows that when using hybrid heat source for cladding, the dilution rate of the cladding layer increases with increasing laser power; the aspect ratio decreases with increasing laser power. As the laser power increases, the penetration depth increases and the height of the cladding layer remains basically unchanged.

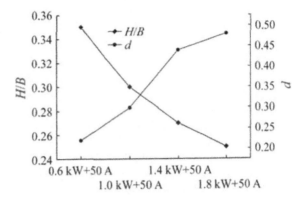

Fig. 5 Effect of different laser powers in hybrid heat source cladding on H/B and d

Therefore, with the increase of laser power, the dilution ratio of the hybrid heat source cladding layer increases, and the aspect ratio decreases.

3.2 The Influence of TIG Current on High-Power Laser-TIG Hybrid Cladding Layer Formation

Figure 6 shows the morphology and cross section of the cladding layer obtained by hybrid heat source when the powder thickness is 1.5 mm, the laser power is 1.8 kW, the cladding speed is 1.5 mm/s, and the TIG current varies from 0 to 80 A. It can be seen from Fig. 6 that to compare 1.8 kW + 0 A and 1.8 kW + 50 A, the former has a significantly smaller weld width and poorer edge formation; with the increase of TIG current, the cladding layer formation becomes better.

Figures 7 and 8 show the changing trend of TIG current for single-clad cladding of laser-TIG composite heat source under the conditions of scanning speed 1.5 mm/s and laser power 1.8 kW. Figure 7 shows that when using hybrid heat source for cladding, the introduction of TIG arc can increase the cladding layer width, but it remains basically unchanged with the increase of TIG current. The laser cladding melting width is basically 5 mm, while the hybrid heat source cladding melting width is basically 7.5 mm. When laser was used alone, the power density reaches a certain value, and the cladding layer width is only affected by the size of the spot and remains basically unchanged. The introduction of TIG arc can increase the melt pool volume, increase the width of the cladding layer, and decrease the wetting angle; the wetting angle remains almost unchanged with the increase of TIG current; the wetting angle θ for single laser cladding is 95°, while the angle for hybrid heat source cladding is 40°–60°.

The introduction of electric arc can increase the energy absorbed by the cladding layer. The high temperature gradient of the edge area can be altered to a mild one, and the stress concentration can be reduced, the spreading ability of the metal powder after melting also can be improved. As the cladding layer width becomes longer, the cladding forming quality was promoted. While the increase of the inputting, more energy can lead to the decrease of the height of the cladding layer. Therefore, the introduction of the TIG arc leads to an increase in the melting width and dilution rate, and a decrease in the wetting angle and aspect ratio. When high-power laser was coupled with a low-current TIG arc for cladding, the energy mainly comes from the laser. With the increase of the TIG current, the energy absorbed by the melt pool, the temperature of the melt pool, and the size of the melt pool gradually increase, so the width and penetration depth of the hybrid heat source cladding layer increase slowly with the increase of the TIG current. The height of the cladding layer remains basically unchanged, so the aspect ratio H/B tends to decrease with the increase of TIG current, and the wetting angle θ decreases with decreasing the aspect ratio; at the same time, as the TIG current increases, the energy absorbed by the edge of the cladding layer will also increase, and the temperature gradient at the edge will

Fig. 6 Morphology and cross-sectional shapes of single cladding with different TGI currents in hybrid heat source cladding

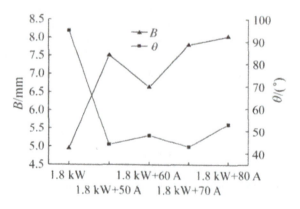

Fig. 7 Effect of different TIG currents in hybrid heat source cladding on B and θ

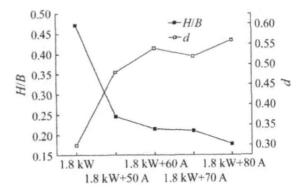

Fig. 8 Effect of different TIG currents in hybrid heat source H/B and d

further decrease. In summary, the wetting angles of the cladding layer obtained by a high-power laser and a low-current TIG arc coupled with a high-power laser are basically the same.

Figure 8 shows that when a high-power laser coupled with a low-current TIG arc for cladding, the dilution rate of the cladding layer remains basically unchanged. However, the aspect ratio decreases through the increase of TIG current. With the increase of the TIG current, the energy absorbed by the powder and the substrate will also increase. As a result, the area of the melt pool increases, and the penetration depth and melting width slowly increase. Under the increase of TIG current, both the height of cladding layer and the dilution rate of the hybrid heat source cladding layer remains basically unchanged, but the aspect ratio of the cladding layer decreases.

4 Conclusion

(1) With more laser power or TIG current, the melting width and dilution rate of cladding layer are increased, and the aspect ratio and the wetting angle of the cladding layer are decreased.
(2) By using a low-current TIG arc in high-power laser overlaying welding process, the aspect ratio, wetting angle, and increase melt width are reduced significantly, and the dilution rate is unchanged. The temperature gradient at the edge of welding pool can also be reduced by TIG arc, and the wetting process can be improved. Also a desired welding appearance is acquired.

Acknowledgements This work was supported by the National Nature Science Foundation of China (51805234), the Program of Innovation Groups of Basic Research of Gansu Province(17JR5RA107), the Foundation of Collaborative Innovation Teams in College of Gansu Province (2017C-07), the Hongliu Excellent Youth Program of Lanzhou University of Technology.

References

1. Zhao MJ, Liang EJ, Zhao D (2003) Effect of TiO_2 on the laser cladding layers of Ni-based alloy on 45 steel. Chin J Lasers 10:947–952
2. Chen B, Yao Y, Tan C et al (2018) Investigation of the correlation between plasma electron temperature and quality of laser additive manufacturing process. Trans Intell Weld Manuf 1:60–74
3. Guo SR, Chen ZJ, Zang QL et al (2013) Research progress on laser surface modification by high-power diode laser. Laser Optoelectron Prog 5:55–62
4. Li DR, Wang FC, Ma Z et al (2006) Mechanics study of thermal shock by laser irradiation in ZrO_2 ceramic coatings. New Technol New Process 6:39–42
5. Xiao FH, Song G (2009) Spectral analysis of plasma in low-power laser/arc hybrid welding of Mg alloy. Trans Plasma Sci 37(1):76–82
6. Li KB, Li D, Liu DY et al (2014) Research of fiber laser cladding repairing process with wire feeding. Chin J Lasers 11:82–87
7. Li X, Shi Y, Zhu M et al (2018) Analysis of spreading of the melt in diode laser-TIG hybrid cladding process. Trans Intell Weld Manuf 1:143–150

Ming Zhu was born in 1984 at Lanzhou Province, China. Mainly engaged in new welding methods and research on welding intelligent control.

Address: School of Materials Science and Engineering, Lanzhou University of Technology, Lanzhou 730050, China.

Information for Authors

Aims and Scopes

Transactions on Intelligent Welding Manufacturing (TIWM) is authorized by Springer for periodical publication of research papers and monograph on intelligentized welding manufacturing (IWM).

The TIWM is a multidisciplinary and interdisciplinary publication series focusing on the development of intelligent modeling, controlling, monitoring, and evaluating and optimizing the welding manufacturing processes related to the following scopes:

- Scientific theory of intelligentized welding manufacturing
- Planning and optimizing of welding techniques
- Virtual and digital welding /additive manufacturing
- Sensing technologies for welding process
- Intelligent control of welding processes and quality
- Knowledge modeling of welding process
- Intelligentized robotic welding technologies
- Intelligentized, digitalized welding equipment
- Telecontrol and network welding technologies
- Intelligentized welding technology applications
- Intelligentized welding workshop implementation
- Other related intelligent manufacturing topics.

Submission

Manuscripts must be submitted electronically in WORD version on online submission system: https://ocs.springer.com/ocs/en/home/TIWM2017. Further assistance can be obtained by emailing Editorial Office of TIWM, Dr. Yan ZHANG: zhangyan521@sjtu.edu.cn, or one of the Editors-in-Chief of TIWM.

Style of Manuscripts

The TIWM includes two types of contributions in scopes aforementioned, the periodical proceedings of research papers and research monographs. Research papers include four types of contributions: Invited Feature Articles, Regular Research Papers, Short Papers and Technical Notes. It is better to limit the full length of Invited Feature Articles in 20 pages; Regular Research Papers in 12 pages; and Short Papers and Technical Notes both in 6 pages. The cover page should contain Paper title, Authors name, Affiliation, Address, Telephone number, Email address of the corresponding author, Abstract (100–200 words), Keywords (3–6 words) and the suggested technical area.

Format of Manuscripts

The manuscripts must be well written in English and should be electronically prepared preferably from the template "splnproc1110.dotm" which can be downloaded from the Web site: http://rwlab.sjtu.edu.cn/tiwm/index.html. The manuscript including texts, figures, tables, references, and appendixes (if any) must be submitted as a single WORD file.

Originality and Copyright

The manuscripts should be original and must not have been submitted simultaneously to any other journals. Authors are responsible for obtaining permission to use drawings, photographs, tables, and other previously published materials. It is the policy of Springer and TIWM to own the copyright of all contributions it publishes and to permit and facilitate appropriate reuses of such published materials by others. To comply with the related copyright law, authors are required to sign a Copyright Transfer Form before publication. This form is supplied to the authors by the editor after papers have been accepted for publication and grants authors and their employers the full rights to reuse of their own works for noncommercial purposes such as classroom teaching.

Author Index

C
Cao, Yue, 95
Chen, Bo, 47
Chen, Chao, 65
Cheng, Han, 47
Chen, Haiping, 83
Chen, Shanben, 65, 83
Chen, Yuhua, 119
Chen, Zhiwei, 47

F
Fan, Ding, 3, 27, 161
Feng, Jicai, 47

G
Gao, Guoxue, 133
Gao, Hui, 133
Ge, Yu, 65

H
He, Yinshui, 109
Huang, Yongde, 119
Hu, Shengsun, 95

L
Li, Gang, 83
Li, Jian, 109
Li, Juan, 133
Li, Shilong, 147
Liu, Yukang, 3
Li, Xubin, 161
Luo, Yu, 133
Lv, Na, 83

M
Ma, Guohong, 109

P
Pang, Yibin, 147

S
Shi, Yu, 3, 27, 161

T
Tan, Caiwang, 47

W
Wang, Long, 133
Wang, Shanlin, 119
Wang, Wandong, 95
Wang, Zhijiang, 95

X
Xin, Jijun, 119
Xu, Yanling, 65

Y
Yan, Buyun, 161
Yu, Huanwei, 65
Yu, Xiaokang, 109

Z
Zhang, Gang, 3, 27
Zhang, Pengxian, 147
Zhang, Yuming, 3
Zhang, Ziyang, 119
Zhou, Canfeng, 133
Zhu, Ming, 27, 161
Zou, Shuangyang, 95

© Springer Nature Singapore Pte Ltd. 2019
S. Chen et al. (eds.), *Transactions on Intelligent Welding Manufacturing*,
Transactions on Intelligent Welding Manufacturing,
https://doi.org/10.1007/978-981-13-7418-0